MATHEMATICIAN'S DELIGHT

W. W. Sawyer

DOVER PUBLICATIONS, INC.
Mineola, New York

Bibliographical Note

This Dover edition, first published in 2007, is an unabridged republication of the 1991 printing of the work originally published by Penguin Books, London, 1943.

Library of Congress Cataloging-in-Publication Data

Sawyer, W. W. (Walter Warwick), 1911–
   Mathematician's delight / W. W. Sawyer.
     p. cm.
   "This Dover edition, first published in 2007, is an unabridged republication of the work originally published in 1991 by Penguin Books, New York.
   Includes bibliographical references and index.
   ISBN-13: 978-0-486-46240-0
   ISBN-10: 0-486-46240-4
   1. Mathematics. I. Title.

QA37.3S28 2007
510—dc22

2007012783

Manufactured in the United States of America
Dover Publications, Inc., 31 East 2nd Street, Mineola, N.Y. 11501

CONTENTS

PART I

THE APPROACH TO MATHEMATICS

CHAPTER 1

THE DREAD OF MATHEMATICS

'The greatest evil is fear.'
Epicurean Philosophy

THE main object of this book is to dispel the fear of mathematics. Many people regard mathematicians as a race apart, possessed of almost supernatural powers. While this is very flattering for successful mathematicians, it is very bad for those who, for one reason or another, are attempting to learn the subject.

Very many students feel that they will never be able to understand mathematics, but that they may learn enough to fool examiners into thinking they do. They are like a messenger who has to repeat a sentence in a language of which he is ignorant – full of anxiety to get the message delivered before memory fails, capable of making the most absurd mistakes in consequence.

It is clear that such study is a waste of time. Mathematical thinking is a tool. There is no point in acquiring it unless you mean to use it. It would be far better to spend time in physical exercise, which would at least promote health of body.

Further, it is extremely bad for human beings to acquire the habit of cowardice in any field. The ideal of mental health is to be ready to face any problem which life may bring – not to rush hastily, with averted eyes, past places where difficulties are found.

Why should such fear of mathematics be felt? Does it lie in the nature of the subject itself? Are great mathematicians essentially different from other people? Or does the fault lie mainly in the methods by which it is taught?

Quite certainly the cause does *not* lie in the nature of the subject itself. The most convincing proof of this is the fact that people in their everyday occupations – when they are making something – do, as a matter of fact, reason along lines *which are essentially the same as those used in mathematics:* but they are unconscious of this fact, and would be appalled if anyone suggested that they should take a course in mathematics. Illustrations of this will be given later.

The fear of mathematics is a tradition handed down from days when the majority of teachers knew little about human nature, and nothing at all about the nature of mathematics itself. What they did teach was an imitation.

Imitation Subjects

Nearly every subject has a shadow, or imitation. It would, I suppose, be quite possible to teach a deaf and dumb child to play the piano. When it played a wrong note, it would see the frown of its teacher, and try again. But it would obviously have no idea of what it was doing, or why anyone should devote hours to such an extraordinary exercise. It would have learnt an imitation of music. And it would fear the piano exactly as most students fear what is supposed to be mathematics.

What is true of music is also true of other subjects. One can learn imitation history – kings and dates, but not the slightest idea of the motives behind it all; imitation literature – stacks of notes of Shakespeare's phrases, and a complete destruction of the power to enjoy Shakespeare. Two students of law once provided a good illustration: one learnt by heart long lists of clauses; the other imagined himself to be a farmer, with wife and children, and he related everything to this farm. If he had to draw up a will, he would say, 'I must not forget to provide for Minnie's education, and something will have to be arranged about that mortgage.' One moved in a world of half-meaningless words; the other lived in the world of real things.

The danger of parrot-learning is illustrated by the famous howler, 'The abdomen contains the stomach and the bowels, which are A, E, I, O and U.' What image was in the mind of the child who wrote this? Large metal letters in the intestines? Or no image at all? Probably it had heard so many incomprehensible statements from the teacher, that the bowels being A, E, I, O and

U seemed no more mysterious than other things heard in school.

A large proportion of examination papers contain mathematical errors which are at least as absurd as this howler, and the reason is the same – words which convey no picture, the lack of realistic thinking.

Parrot-learning always involves this danger. The deaf child at the piano, whatever discord it may produce, remains unaware of it. Real education makes howlers impossible, but this is the least of its advantages. Much more important is the saving of unnecessary strain, the achievement of security and confidence in mind. It is far easier to learn the real subject properly, than to learn the imitation badly. And the real subject is interesting. So long as a subject seems dull, you can be sure that you are approaching it from the wrong angle. All discoveries, all great achievements, have been made by men who delighted in their work. And these men were normal, they were no freaks or high-brows. Edison felt compelled to make scientific experiments in just the same way that other boys feel compelled to mess about with motor bicycles or to make wireless sets. It is easy to see this in the case of great scientists, great engineers, great explorers. But it is equally true of all other subjects.

To master anything – from football to relativity – requires effort. But it does *not* require *unpleasant* effort, drudgery. The main task of any teacher is to make a subject interesting. If a child left school at ten, knowing nothing of detailed information, but knowing the pleasure that comes from agreeable music, from reading, from making things, from finding things out, it would be better off than a man who left university at twenty-two, full of facts but without any desire to inquire further into such dry domains. Right at the beginning of any course there should be painted a vivid picture of the benefits that can be expected from mastering the subject, and at every step there should be some appeal to curiosity or to interest which will make that step worth while.

Bad teaching is almost entirely responsible for the dislike which is shown in such words as 'high-brow'. Children want to know things, they want to do things. Teachers do not have to put life into them: the life is there, waiting for an outlet. All that is needed is to preserve and to direct its flow.

Too often, unfortunately, teaching seems to proceed on the philosophy that adults have to do dull jobs, and that children should get used to dull work as quickly as possible. The result is an entirely justified hatred and contempt for all kinds of learning and intellectual life.

Many members of the teaching profession are already in revolt against the tradition of dull education. Some excellent teaching has been heard over the wireless. The same ideas, the same methods are being developed independently in all parts of the country. No claim for originality is therefore made in respect of this book. It is no more than an individual expression of a feeling shared by thousands.

In the following chapters I shall try to show what mathematics is about, how mathematicians think, when mathematics can be of some use. In such a short space it is impossible to go into details. If you want to master any special department of mathematics, you will certainly need text-books. But most text-books contain vast masses of information, the object of which is not always obvious. It would be useless to burden your memory with all this purposeless information. It would be like having a hammer so heavy that you could not lift it. Mathematics is like a chest of tools: before studying the tools in detail, a good workman should know the object of each, when it is used, how it is used, what it is used for.

CHAPTER 2

GEOMETRY – THE SCIENCE OF FURNITURE
AND WALLS

'So the Doctor buckled to his task again with renewed energy; to Euclid, Latin, grammar and fractions. Sam's good memory enabled him to make light of the grammar, and the fractions too were no *great difficulty, but the Euclid was an awful trial. He could not make out what it was all about. He got on very well until he came nearly to the end of the first book and then getting among the parallelogram 'props' as we used to call them (may their fathers' graves be defiled!) he stuck dead. For a whole evening did he pore patiently over one of them till* AB, *setting to* CD, *crossed hands, poussetted and whirled round 'in Sahara waltz' through his throbbing head. Bed-time, but no rest! Who could sleep with that long-bodied ill-tempered looking parallelogram* AH *standing on the bedclothes, and crying out in tones loud enough to waken the house, that it never had been, nor ever would be equal to the fat jolly square* CK?'*

                    Henry Kingsley, *Geoffrey Hamlyn.*

IN the previous chapter it was mentioned that people, in their everyday life, used the same methods of reasoning as mathematicians, but that they did not realize this.

    For instance, many people who would be paralysed if you said to them, 'Kindly explain to me the geometrical construction for a rectangle' would have no difficulty at all if you said, 'Please tell me a good way to make a table.' A 'rectangle' means the shape below –

and no one could make much of a table unless he understood well what this shape was. Suppose for instance you had a table like this

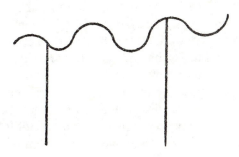

All the plates and tea-pots and milk-jugs would slide down into the hollows, or fall over, and altogether it would be very inconvenient. People who make tables are unanimous that the tops ought to be *straight*, not curved. Even if the top is straight, it may not be level; the table may look like this ⟋⟨. And when the top is right, the legs may still look queer, such as ⟋ ⟋ or ⟋⟨. In such cases the weight of the table-top would tend to break the joints. To avoid this, legs are usually made upright, and the table stands on the floor like this ⎿ ⏌.

Anyone who understands what a table should look like understands what a rectangle is. You will find a lot about rectangles in books on geometry, because this shape is so important in practical life – though the older geometry books give no hint of this reason *why* we study rectangles.

Another craft which uses rectangles is bricklaying. An ordinary brick has a rectangle on top, below, at the ends and sides. Why? It is easy to guess. The bricks have to be laid level, if they are not to slide. (Even in making walls from rough stone, such as the Yorkshire dry walls, one tries to build with level layers.) So that the bricks must fit in between two level lines. But it would still be possible to have fancy shapes for the ends –

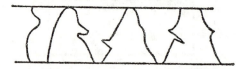

But this looks more like a jig-saw puzzle than a wall: the poor bricklayer would spend half his life looking for a brick that would fit. We want all the bricks to have the same shape. This can be done in several ways – ⫽⫽⫽⫽⫽⫽ or ⫞⫞⫞⫞⫞. These would make ragged ends to the wall, and if two walls met there would be open spaces to fill. By having the ordinary shape of brick, all these complications are avoided.

No one will have any difficulty in following such an argument. Why, then, do people dislike geometry? Partly because it is a mystery to them: they do not realize (and are not told) how close it lies to everyday life. Secondly, because mathematics is supposed to be *perfect*. There is nothing in the geometry book about shapes being 'nearly triangles' or 'almost rectangles', while it is quite common for a door or table to be just a little out of true. This perfection puts people off. You can have several tries at making a table, and each attempt may be an improvement on the last. You learn as you go along. By insisting on 'mathematical exactness', it is easy to close this great road of advance, Trial and Error. If you remember how close geometry is to carpentry, you will not fall into this mistake. If you have a problem which puzzles you, the first thing to do is to try a few experiments: when you have found a method that seems to work, you may be able to find a logical, 'exact', 'perfect' justification for your method: you may be able to prove that it is right. But this perfection comes at the *end*: experiment comes at the beginning.

The first mathematicians, then, were practical men, carpenters and builders. This fact has left its mark on the very words used in the subject. What is a 'straight line'? If you look up 'straight' in the dictionary, you will find that it comes from the Old English word for 'stretched', while 'line' is the same word as 'linen', or 'linen thread'. A straight line, then, is a stretched linen thread – as anyone who is digging potatoes or laying bricks knows.

Euclid puts it rather differently. He says a straight line is the shortest distance between two points. But how do you find the shortest distance? If you take a tape-measure from one point to the other, and then pull one end as hard as you can, so that as little as possible of the tape-measure is left between the two points, you will have found the shortest path from one to the other. And the tape-measure will be 'stretched' in exactly the same way as the builder's or gardener's 'line'.

If you are told to define something, ask yourself, 'How would I make such a thing in practice?'

For instance, you might be asked to define a 'right angle'. A 'right angle' (in case the expression is new to you) means the figure formed when two lines meet as in a capital L, thus: |_ . You will find a right angle at every corner of this sheet of paper. On the other hand,/_ and __/ are not right angles.

How would you make a right angle? Suppose you want to tear

a sheet of note-paper into two neat halves: what do you do? You fold it over, and tear along the crease, which you know stands at the 'right' angle to the edge. If you fold it very carelessly, you do not get the 'right' angle, but something like_/__ : too much paper is left on one side, too little on the other. We now see the special feature of a right angle – both sides of the crease *look the same*. If we had blots of ink on one side of the crease, we should get 'reflections' of these on the other when we unfolded the paper. The crease acts like a mirror. And the reflection of the edge of the paper – if we have the right angle – lies along the edge on the other side of the crease.

You can try this with a ruler or walking-stick. There is a position in which a stick can be held so that its reflection seems to be a continuation of the stick: you can look along the stick and its reflection, exactly as if you were squinting down the barrel of a rifle. The stick is then 'at right angles' to the mirror.

But suppose you are laying out a football field, and want to get a right angle. You cannot fold the touch-line over on to itself and notice where the crease comes! But this idea of a mirror shows a way of getting round the difficulty.

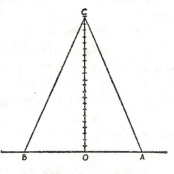

Suppose O is the point on the touch-line where you want to draw a line at right angles to the touch-line, OA. We know that a mirror, OC, in the correct position, would reflect the point A so that it appeared at B, also on the touch-line. If we folded the paper over the line OC, A would come on top of B. The line OA would just cover OB, and the line AC would just cover BC, after such folding.

But this suggests a way of finding the line OC. If we start at O, and measure OA, OB the same distance on opposite sides of O, we have A and its reflection, B. Since BC is the reflection of AC, both must be the same length. Take a rope of any convenient length, fasten one end to A, and walk round, scraping the other end of the rope in the ground. All the points on this 'scrape' will be a rope's length from A. Untie the rope from A, and fix it to B instead, and make another, similar scrape in the ground. Where the two scrapes cross, we have a point which is the same distance from B as it is from A. This will do for C. We drive a peg in

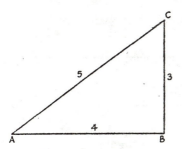

here, stretch a line from C to O, and whitewash along it.

You can easily see how the above method, suitable for football fields, can be translated into a method for drawing right angles on paper with ruler and compass.

But there is another, very remarkable, way, which is actually used for marking out football fields.

If you take three rods of lengths 3, 4, and 5 yards, and fit them together as shown in the figure, you will find that the angle at B turns out to be a right angle. No one could have guessed that this would be so. It seems to have been discovered about five thousand years ago, more or less by accident. It is not known who discovered it, but the discoverer was almost certainly someone engaged in the building trade – a workman or an architect. This way of making a right angle was used as part of the builder's craft: people did not ask why it was so, any more than a housewife asks why you use baking-powder. It was just known that you got good results if you used this method, and the Egyptians used it to make temples and pyramids with great success.

It is not known how far learned Egyptians bothered their heads trying to find an explanation of this fact, but certainly Greek travellers, who visited Egypt, found it a very intriguing and mysterious thing. Egyptian workmen saw nothing remarkable in it: if the Greeks asked them about it, they probably answered, 'Lor' bless you, it's always been done that way. How else would you do it?'

So the Greeks would go away still wondering, 'Why?' Why

3, 4, and 5? Why not 7, 8, and 9? Anyhow, what does happen if
you try 7, 8, and 9? Or any other three numbers?

It would therefore be quite natural to start with fairly small
numbers, and try making triangles, such as (1, 1, 1), (1, 1, 2),
(1, 1, 3), (1, 2, 2), (2, 2, 2), etc. The Greeks had no Meccano –
with Meccano it is easy to make such triangles quickly. How do
they look?

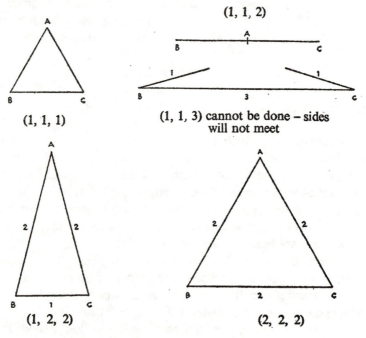

As soon as you start experimenting in this way, you begin to
discover things. You sometimes find that it is impossible to make
the triangle at all; e.g., (1, 1, 3), (1, 1, 4) and so on: in fact whenever one side (e. g., 3) is bigger than the other two sides (1 and 1)
put together.

You may notice that doubling the sides of a triangle does not
alter its shape: (2, 2, 2) looks much like (1, 1, 1).

Again the triangle (1, 2, 2) has a pleasing balanced appearance:
if you turned it over, so that B and C changed places, it would
still look just the same.

The more you experimented with drawing or making triangles, the more things you would notice about them. Not all these discoveries would be really new. For instance, we saw above that, in any triangle, AB plus AC must be bigger than BC. But this is not new. We know that the straight line BC is the shortest way from B to C, so naturally it is longer if we go from B to C via A, a distance equal to the sum of AB and AC. So that this particular result *could* have been found by reasoning: it follows from the fact that the straight line gives the shortest path between two points.

So that we can do two things in our study of the shapes of things. (i) We can discover a large number of facts. (ii) We can arrange them in a system, showing what follows from what.

Actually, these two things were done by the Greeks, and by 300 B.C. Euclid had written his famous book on geometry, putting all the facts known into the form of a system. In this book you will find why (3, 4, 5) gives a right-angled triangle; and it is shown that other triangles, such as (5, 12, 13) or (24, 25, 7) or (33, 56, 65), do the same.

But all this took time. The Great Pyramid was built in 3900 B.C., by rules based on practical experience: Euclid's system did not appear until *3,600 years later*. It is quite unfair to expect children to start studying geometry in the form that Euclid gave it. One cannot leap 3,600 years of human effort so lightly! The best way to learn geometry is to follow the road which the human race originally followed: *Do* things, *make* things, *notice* things, *arrange* things, and only then – *reason* about things.

Above all, do not try to hurry. Mathematics, as you can see, does not advance rapidly. The important thing is to be sure that you know what you are talking about: to have a clear picture in your mind. Keep turning things over in your mind until you have a vivid realization of each idea. Once you have learnt how to think in clear pictures, you will advance quickly, without strain. But it is fatal to advance and to leave the enemy – confused thought – in your rear. Rather than this, start again at the multiplication tables!

SOME EXPERIMENTS CONNECTED WITH GEOMETRY

1. A boy has a strip of wood, AC, which is 4 feet long. He wishes to join a second strip to it, as in the figure, so that a string, ABCD, passed round the outside, will form a rectangle.

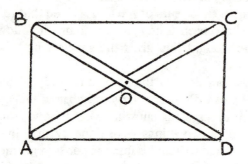

How long must he make the strip BD? At what point (O) must he nail the two strips together? Does it matter at what angle he places the two strips?

2. A flat piece of ground is to be covered with tiles. All the tiles must be the same shape and size, but it does not matter if there is a jagged edge on the outside border. Design as many different ways of doing this as you can. One example is shown in the drawing below.

3. A street lamp is 12 feet above the ground. A child, 3 feet in height, amuses itself by walking in such a way that the shadow of its head moves along lines chalked on the ground. How will it

walk if the chalked line is (i) straight, (ii) a circle, (iii) a square? What is the rule connecting the size and shape of the child's track with those of the chalked line? (*Note* – Do not quarrel with anyone about the answer to this, until you have actually made an experiment. A convenient form of experiment is to take an indoor lamp instead of the street lamp, and a pencil to represent the child. The pencil will record its own track as it moves.)

4. What difference would it make to the last question if the light came from the sun instead of from a lamp?

5. A man 6 feet tall stands at a distance of 10 feet from a lamppost. The lamp is 12 feet above the ground. How long will the man's shadow be?

6. A hiker can see two church spires. One is straight in front of him. The other is directly to the left of him. He has a map on which the two churches are marked, but he has not the least idea of the direction in which he is facing, whether it is North or South, or any other point of the compass. What can he tell about where he is on the map? (Suggested method. Drive two nails into a flat piece of wood. Let these represent the churches. Take a piece of cardboard, one corner of which is a right angle, and fit this between the nails, as in the figure. Then P is a possible position for the hiker. For if he is

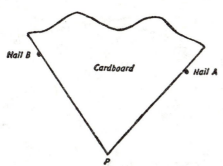

looking in the direction PA, B will be directly on his left. Mark the position P on the wood. Slide the cardboard, and mark other possible positions in the same way. These marks all lie on a certain curve. What is the curve?)

7. In a miniature rifle-range, 25 yards long, it is desired to construct a moving target, to represent a lorry, 20 feet long and 15 feet high, half-a-mile away, moving at 20 miles an hour. The marksman is supposed to be in such a position that he sees one side of the lorry. How large should the model be, and how quickly should it move across the screen?

8. A spider wishes to crawl from one corner of a brick, A, to the opposite corner, B, by the shortest possible way. Which path should it take? The spider of course crawls over the surface of the brick – it cannot burrow through the brick.

(Material required – a number of bricks, having different shapes, and a piece of string to stretch from A to B. It is useful to make the bricks out of cardboard, by folding. After the

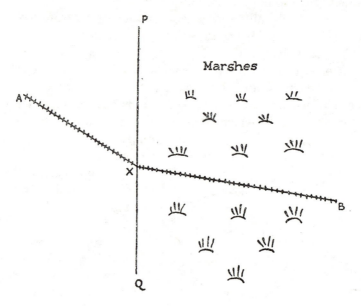

A to X, £10,000 a mile.   X to B, £20,000 a mile.
Find the best position for X.

shortest path has been found and marked on the cardboard, the cardboard can again be flattened out, and the form the path then takes should be noted.)

9. Get a globe of the world. Stretch a thread between two places. Make a note of the places over which the thread passes. Mark these places on a map of the world in an atlas (Mercator's projection). Notice how different the curve joining them is from a straight line drawn on the map. This fact is important for sailors, and airmen flying long distances ('Great Circle Navigation').

10. A railway is to be built joining two towns, A and B. The ground to the right of the line PQ is marshy, and as a result it

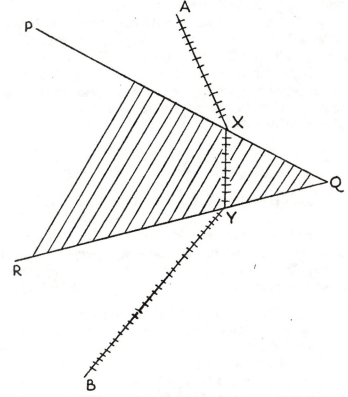

A to X £10,000 a mile. X to Y, £20,000 a mile. Y to B, £10,000 a mile.
Find the best positions for X and Y.

costs twice as much to build one mile of railway here as it costs to build one mile on the ground to the left of PQ. Draw a number of possible routes for the railway from A to B, work out the costs of construction, and find, as nearly as you can, the route which makes the construction as cheap as possible. (See Fig., p. 20.)

(*Note* – You are *not* expected to answer this question purely by calculating. Draw a plan for yourself, put the towns A and B wherever you like, measure the lengths of any lines you want, assume that one mile of railway on the left of PQ costs £10,000. In real life, we want to get the answer by hook or by crook – by calculation, or by experiment, or by a mixture of the two. Never mind what Euclid said in 300 B.C.)

11. This question is similar to the last, except that the shape of the obstacle is different. A railway has to be built between A and B, but a wedge of difficult ground, PQR, lies between them. Find the best route for the railway. (This type of problem does occur in practice, when hilly country lies between towns. In this case, the extra expense would be due to the need to excavate cuttings.) (See Fig., p. 21.)

CHAPTER 3

THE NATURE OF REASONING

'*From my observations of men and boys I am inclined to think that my way of study is the common way, the natural way, and that the schoolmasters destroy it and replace it by something that conduces to mere learning.*'   John Perry, 1901.

BERNARD SHAW once made an unkind remark to the effect that people who knew how to do anything went and did it; while those who did not know how to do anything were obliged to earn a living by teaching.

Actually, teaching is far harder than doing. You will find a hundred men who are brilliant footballers for every one that can teach you how to play the game well. You will find hundreds of clever children and hundreds of dull children, but you rarely find a child who was dull to begin with but became clever through the help of a teacher. But that is the test of a teacher. Most teachers who are honest with themselves are forced to admit that, in the main, the class would make the same progress if the teacher were not there at all, the clever remaining clever and the dull remaining dull.

Are there in fact two races of men, those who are born to succeed and those who are born to fail? Have the 'great men' some special way of thinking which ordinary people lack, and cannot understand?

There are, of course, certain differences between the bodies and brains which children inherit from their parents. There are cases of mental deficiency, where important glands fail to do their job, and the children have to be kept in special homes. It may be that glands, or other factors, set a natural limit to the powers of each one of us, and that it is foolish to strive for skill beyond this limit. This may be so. It is certain that not one person in a thousand makes full use of the glands and the brains actually possessed, or comes anywhere near the natural limit of intelligence. We certainly cannot explain it by glands if a person is bright and resourceful outside the classroom, and dull in everything connected with school. The reason must be sought elsewhere.

It is extremely interesting to inquire just what it is that the 'great man' or the successful performer does, that others fail to do. Just what qualities are necessary in order to play a game well, to be a painter, a musician, an engineer, a farmer or a mathematician? Can these qualities be developed by suitable exercises? Is it possible for an ordinary person, with determination, to acquire these qualities? When teachers are in a position to answer these questions, when everybody reaches the limit of his powers, the time will have come to talk about inherited differences in intelligence. But that will be a few centuries hence.

At present there are books which do really teach. It is better to spend hours searching a big library for such a book, rather than to read hundreds of books by second-rate authors. It is very unlikely that you or I possess any really original ideas. People seem to belong to certain types. If you feel strongly on any subject, the chances are that you will find some other individual has been concerned with exactly the same question, and you will find your own views worked out in his or her writings. You can then start to study the subject where that earlier worker left off.

One can often get help in teaching or learning a subject by reading books written on other subjects. In a library I came across a book, *Swimming for All*, by R. C. Venner (G. Bell & Sons, 1933). This book follows a method which could probably be applied to most other subjects. First the author explains the principles of swimming. He explains the difference between the movements which are needed in the water and those which we instinctively make as a result of living on dry land. Then he gives a series of experiments and exercises, by which one can convince oneself of the truth of his remarks, so that one not merely knows these facts, but comes to *feel* their truth and to do the right thing instinctively.

Writers on tennis make a remark which may serve as a parable. They say that you should not begin by trying to strike the ball into the court, but that you should begin by hitting it hard, and with a good style. Gradually you will find that the ball begins to land in the court. If you start by worrying about where the ball goes, you will always be a feeble player. Much the same holds in mathematics. The important thing is to learn how to strike out for yourself. Any mistakes you make can be corrected later. If you start by trying to be perfect, you will get nowhere. The road to perfection is by way of making mistakes.

This rather reminds me of a book on drawing, which I read several years ago. Unfortunately I did not make a note of the author's name.* He tried to teach drawing in such a way that his readers would be able to sit on the top of an omnibus and to record on paper the fleeting expressions of people's faces. According to him, you should use odd scraps of paper, and never alter a drawing. Throw it away if it goes wrong, and begin again. Do not bother if things are the right shape or not. Jot down

*Probably *Drawing for Children* by Vernon Blake. One remark quoted, however, is from L. Doust, *How to Sketch from Life*.

what you actually see, especially shadows. Do not draw lines unless you actually see them standing out. Regard your first drawings merely as sketches, noting what you actually see. Gradually you will find that the shapes of your drawings become more true to life, but even your earlier drawings, which have the wrong proportions, will suggest something solid and real. He gave very rough sketches to illustrate this fact. I know nothing about drawing, but if I wanted to learn, I should certainly learn this way.

In all subjects, it seems, there is a way of approach which is both interesting and encouraging. The 'great men' are often those who felt a strong interest in a subject and by accident, by experiment or through the influence of good teachers, hit on the correct approach. It is *ignorance of the way in which a subject is tackled* that causes geniuses to be regarded as a race apart. The more one studies the methods of the great, the more commonplace do these methods appear.

Very often anecdotes give us a false impression. There is a story about Newton and an apple: Newton saw an apple fall, and wondered *why* it fell – so we are told. It is extremely unlikely that Newton did anything of the sort. To this day we do not know *why* an apple falls. It is more likely that Newton thought rather as follows. What would happen if an apple were dropped from a very great height? Presumably it would still fall, however high, however far away from the earth it was taken. If not, there would be some height at which you suddenly found that the apple did not fall. This is possible, but not very likely. It seems probable, then, that, even if you went as far as the moon or the sun, you would still feel the pull of the earth, though perhaps not so strongly as you do here. Perhaps it is this pull which keeps the moon close to the earth, and keeps the earth circling round the sun? That at any rate was the conclusion to which the apple led Newton – that every piece of matter in the universe exerts a pull on every other piece, however far away it may be. No one denies the greatness of Newton. Equally, no one can say that there was anything superhuman about the type of argument he used.

Reasoning in Mathematics

Mathematics teaches us how to solve puzzles. Everyone knows that it is easy to do a puzzle if someone has told you the answer. That is simply a test of memory. You can claim to be a mathematician if, and only if, you feel that you will be able to solve a puzzle that neither you, nor anyone else, has studied before. That is the test of reasoning.

What exactly is this power of reasoning? Is it something separate from the other powers of our minds? Is it something fixed, or something that can be trained and encouraged? How do we come to possess such a power?

Mathematical reasoning does at first sight seem to be in a class by itself. It seems to find a place neither in the experimental sciences, nor in the creative arts.

Some subjects are clearly the result of experiment, or of experience. Chemistry deals with what happens when metals are dropped into liquids, or when the contents of one pot are mixed with those of another. Mechanics deals with the motion of solid objects. History records the actions of men. The study of languages deals with the words used by nations in different parts of the world. It is easy to see how the information contained in a book on chemistry, mechanics, history or French is obtained.

On the other hand there are the subjects (beloved by some people on which everybody disagrees. These are the subjects which do not depend on evidence at all – what you like, what you think ought to be done, the kind of person you admire, the political party you vote for: these are things for which you yourself take responsibility, they show what sort of person you are. You may be ready to fight to secure the type of world you think best: indeed, you should be. But you do not change your basic ideas of what is desirable as the result of argument and evidence. I suppose microbes have a vision of a world made safe for smallpox. We cannot *prove* that the world was not made for the benefit of microbes. All we can do is to use plenty of disinfectant.

Mathematics seems a peculiar subject. It is not a matter of taste. In it, more than in any other science, there is an answer which is right and an answer which is wrong. But, on the other hand, it does not seem to deal with anything definite. A great and important part of mathematics, for instance, deals with the square root of minus one – something which no one has ever seen or felt or tasted. Yet there is no sort of doubt about its properties.

In ancient times philosophers found it hard to explain man's powers of reasoning, and were led to more or less fantastic explanations. One such theory was that we lived in another world before we were born, and in that world were acquainted with the laws of arithmetic and geometry (how far the syllabus went I do not know). The object of education, in this world, was simply to awaken in us the memory of this knowledge.

One should not sneer at this ancient theory. It at least makes clear that education consists in co-operating with what is already inside a child's mind. A good teacher will nearly always be able to make his points simply by asking a class questions, by making the class realize clearly what they already know 'at the back of their minds'.

The clue which we possess today, which the old philosophers could hardly have guessed, comes from biology. It is now generally accepted that life has been on the earth for millions of years, and that we are born with instincts tested and tried in an age-long struggle for survival. On top of these instincts we have a training, given to us especially in the first five years of life, and based on traditions, some of which go back to the experience of thousands of years ago. By the time we are five years old we are, so to speak, a highly manufactured article, and it is generally after this age that we become aware of our ability to argue things out for ourselves.

If therefore we find in ourselves a strong desire to do a certain thing, or to believe a certain thing, it is at least possible that this desire exists in man, because it has enabled him to survive, because it has proved its worth in generations of struggle with the actual world. Those animals, those races survive, which *do* the correct thing. We should therefore expect human brains, human minds to be made on the whole in such a way as to produce *correct action* in any situation. We should *not* expect them to be made in such a way as to produce perfect, logical thought. In fact we often do find people doing the right thing for the wrong reason. A certain place is a centre of infection: savages know nothing of the causes of disease: they say that it is unlucky to go there, it is the abode of an evil spirit.

It was shown in Chapter 2 that geometry did, as a matter of fact, pass through the stage in which workmen *did* the right thing, but had no theory to explain why. Both geometry and arithmetic are closely connected with everyday life – geometry with building, arithmetic with money payment. If you give a tram-conductor threepence for two twopenny tickets, he is not prepared to believe that twice two being four is purely a university fad. He regards it as a fact, well established by the experience of everyday life.

It is now beginning to seem that mathematics is, like chemistry, something that we learn through experience of the real world. Some people will object strongly to this. They will say, 'I can imagine zinc being dropped into sulphuric acid and nothing happening. But can you imagine twice two being five?'

I certainly cannot imagine twice two being five. If a man claimed to perform miracles, and could turn twice two into five, I should give him full marks. It would impress me far more than any other miracle.

But that is not the point. The point is, *why* can we not imagine two twos making five?

There are two possible explanations. (i) We possess some mysterious faculty, given us in a previous life, or by other means. (ii) We cannot imagine twice two being five because in the whole of human history, twice two always has been four, and there has been no need for our minds to imagine it any other way.

The first explanation does not agree with the experience of most teachers. There may be people who possess a faculty of reasoning so perfect as to support this view: it would be interesting to know which schools and colleges they attend. In the works of the greatest mathematicians one finds evidence of blunders, of misunderstanding, of painful groping towards the truth.

Perhaps the greatest blow to the 'mysterious faculty' theory is the fact that mathematicians of today reject as untrue just those beliefs which earlier mathematicians most firmly believed. It was a custom at one time to say that some doctrine was true as certainly as the angles of a triangle added up to two right angles. If Einstein is right, the angles of a triangle do *not* add up to two right angles. Both the theory of relativity and quantum mechanics have destroyed long-cherished beliefs, and forced us to examine again the foundations of our belief.

If you accept Euclid's geometry because it agrees with what you can see of the shapes of things, you will not be unduly alarmed if someone suggests that Euclid may be wrong by a few millionths of an inch in certain places. For you cannot see a millionth of an inch, and Einstein's geometry only differs from Euclid's by millionths. But if you believe that Euclid represents absolute truth – then you are in a mess. Actually, Euclid himself only said, '*If* you admit certain things, then you must admit that the angles of a triangle add up to two right angles.'

On the other hand, there are interesting signs of the way in which human thought has been built up through daily experience. One such sign is to be found in the words we use. Try to imagine, if you can, a cave-man (or whoever it was that first developed language) trying to say to a friend, 'What this writer says about the square root of minus one does not agree with my philosophy at all.' How would he manage to make his friend understand what he meant by such abstract words as 'philosophy', 'minus one', 'agree', and so forth? Every child, in learning to speak, is faced by the same problem. How does it ever come to know the meaning of words, apart from the names of people and objects it can see?

It is instructive to take a dictionary, and to look up such words. Almost always, one finds that abstract words, the names of things which cannot be seen, come from words for actual objects or actions. Take, for instance, the word 'understand'. Both in German and English it is connected with the words 'to stand under'. In French, 'do you understand?' is 'comprenez-vous?', which means 'Can you take hold of that?' rather like the English phrase, 'Can you grasp that?' Still today, people make such remarks as 'Try to get that into your head.'

In learning to speak, a child follows much the same road. It learns the names of its parents and of household objects. It also learns words which describe its feelings, 'Are you hungry?' 'Are you tired?' 'He looks happy.' 'Don't be frightened.' 'Can't you remember?' 'Say you are sorry.'

Every philosopher, every professor, every school-teacher that ever lived began in this way – with words to describe things seen, or things felt. *And all the complicated ideas that have ever been thought of, rest upon this foundation.* Every writer or speaker that ever invented a new word had to explain its meaning by means of other words which people already knew and understood. It would be possible to draw a huge figure representing the English language, in which each word was represented by a block, resting on other blocks – the words used to explain it. At the bottom we should have blocks which did not rest on anything. These would be the words which we can understand directly from our own experience – what we see, what we feel, what we do.

For example, *philosophy*. Philosophy is what a *philosopher* does. Philosopher means '*a lover of wisdom*'. The meaning of *love*, and of being *wise*, we have to learn from everyday life.

What is true of philosophy is equally true of mathematics: its roots lie in the common experiences of daily life. If you can trace the way in which mathematical terms were gradually developed from everyday words, you can understand what mathematics is.

The essential point to grasp is that *mathematical reasoning* is not separate from the other powers of the mind, nor is mathematics separate from the rest of life. Quite the opposite: mathematics has grown from the rest of life, and reasoning has grown from experience.

Another sign of the way our minds are made is to be found in the law, well known to psychologists, 'There is nothing in imagination which was not previously in sense.' For instance, try to imagine a new colour. You will find that you are simply combining the effect of colours that you have already seen. Or try to imagine heaven, or a perfect world. You will find yourself putting together memories of your happiest moments, or turning upside down the things which have aroused your indignation. In a Scottish school, the children (sitting on hard seats) wrote an essay on 'The Perfect School'. Ninety per cent started by stating that the perfect school had cushions on the seats: they then described how the teachers were kept in fear and trembling by the stern rule of the pupils.

Reasoning and Imagination

Earlier we considered the argument, 'Twice two must be four, because we cannot imagine it otherwise.' This argument brings out clearly the connexion between reason and imagination: reason is in fact neither more nor less than *an experiment carried out in the imagination.* In any good detective story, the detective tries to imagine as clearly as possible the background of a crime, and to see how the statements of various witnesses fit into the picture. We are able to follow the story and the reasoning simply by using our own imagination. (*The Mystery of Marie Roget,* by Edgar Allan Poe, is a good example of imaginative reasoning applied to a crime in real life.)*

It is by no means necessary that reasoning should proceed by clearly stated steps. If you hear a rumour which means that your friend Smith has been involved in some particularly dirty business, you may say, 'I do not believe this story. Smith would never do a thing like that.' You may not be able to quote stories of heroic deeds performed by Smith, or to give evidence of any definite kind at all. You just feel that Smith is a decent person. Yet this is a perfectly good example of reasoning. Whether your conclusion is correct or not will depend on how long, and how well, you have actually known Smith. You will find it very hard to make the public share your faith in Smith. They have not your experience of Smith: therefore they cannot imagine him as you do: therefore they reason differently about him.

It is said that European explorers who told tropical peoples of

*See notes and introduction to Dorothy L. Sayers, *Great Short Stories of Detection, Mystery and Horror*, Part I (Gollancz).

the northern winter, when water became like a stone and men could walk on it, were met with polite disbelief. The natives looked at the warm sea waves rolling beneath the palm-trees, and refused to believe in ice. It was outside their experience: they were familiar with the tales of travellers.

People often make mistakes when they reason about things they have never seen. Children imagine kings wearing crowns; in real life, the odds are that a king wears a military cap or a bowler hat. Before the first locomotives were made, people refused to believe they would work. It was thought that the wheels would slip, and the train would remain motionless. A certain Mr Blenkinsop went so far as to invent a locomotive with spiked wheels to overcome this purely imaginary difficulty.

If anyone in the year 1700 had prophesied what the world would be like today, he would surely have been considered mad.

Imagination does not always give us the correct answer. We can only argue correctly about things of which we have experience or which are reasonably like the things we know well. If our reasoning leads us to an untrue conclusion, the fault lies with our reasoning. We must revise the picture in our minds, and learn to imagine things as they are.

When we find ourselves unable to reason (as one often does when presented with, say, a problem in algebra) it is because our imagination is not touched. One can begin to reason only when a clear picture has been formed in the imagination. Bad teaching is teaching which presents an endless procession of meaningless signs, words and rules, and fails to arouse the imagination.

The main aim of this book is not to explain how problems are solved: it is to show what the problems of mathematics *are*.

Abstraction

Let us now consider an example of reasoning applied to objects which everyone has seen and can therefore imagine correctly. Two railway stations, A and B, are connected by a single-track line. Owing to some mistake, a train leaves A for B at the same time as another train leaves B for A. There are no signals or safety devices on the line. Apart from exceptional happenings (such as a storm tearing up the railway track) one expects that there will be a collision.

You will agree that you could have reached this conclusion by the use of your own imagination. But now notice what a dim picture your imagination gave you. In what sort of country did you imagine the line to lie? Among woods, or through towns, or

1.

2.

STAGES OF ABSTRACTION
1. The rough impression of a scene as it might exist in a person's imagination.

2. A diagram, which leaves out all details except those needed for a particular purpose – namely, to show that the two trains are about to collide.

Most diagrams in mathematics are like '2'. All the details except those needed for a particular purpose are left out. But behind every diagram is a picture, like '1'. If you can discover what the picture is, you will find the diagram much easier to understand.

on the top of a precipice? Did you see clearly in your mind the movements of the pistons and the crowd of little devices on the wheels of the engines? How did you imagine the expression on the faces of the drivers, the colour of their hair, the build of their bodies? Did you imagine goods trains or passenger trains? One could continue thus for ever. It is certain, however vivid your imagination, there were points which you overlooked. But this did not in the least affect your answer to the question: will there be a collision? If you had thought of the two trains as two beads threaded on a wire (and the movement of trains might well be shown in this way when a timetable was being planned), you would still come to the correct conclusion. *For the purpose of this question* the trains and the railway-line might just as well be two beads and a wire. Of course for other purposes – if you had to provide ambulances for the wounded, or if you wanted to paint a picture of the event – it would be necessary to know further details.

It is impossible to imagine any event in perfect detail. In attacking any problem, we simplify the situation to a certain extent. We do not bother about those facts which seem unimportant. The result of our reasoning will be correct if the picture in our imagination is, *not* exactly correct, but *sufficiently correct for the purpose in hand.*

This process of forgetting unimportant details is known as *abstraction.* Without abstraction, thought is impossible. We should spend all our lives collecting information if we tried to make a *perfect* picture even of a simple event. Some mis-educated people continually interrupt sensible discussion by wailing, 'But you have not defined exactly what you mean by this word.' The great majority of words cannot be defined exactly (for instance, the word *red.*) The important thing is not exact definition: it is to know what you are talking about.

Serious misunderstanding can arise if one forgets the nature of abstraction, and tries to apply a picture of the world, which is entirely sufficient for some purpose, to another purpose for which it is entirely insufficient. Two examples of difficulties which arise in this connexion may be mentioned.

The Mechanical View of Life

At one time there was a great craze for explaining everything in terms of machinery. It had been discovered that many facts of nature, in particular the movements of the planets, the tides, and of solid objects on the earth's surface, could be explained by supposing the universe to be made up of hard little balls, attracting each other according to certain definite laws. Instead of saying, 'We have a theory sufficiently correct for certain purposes', philosophers and scientists leapt to the conclusion that they had the whole truth about the universe. Not only the sun and moon, but our brains also, were made out of these hard little balls, and everything we did was a consequence of the way they pulled each other about. Thought and feeling must therefore be pure illusions – this in spite of the fact that the theory itself was the result of thought!

The whole procedure was entirely unscientific. It is obvious to anyone that courage, loyalty, determination, affection are *facts*, just as much as pound weights or spring balances. Without these qualities, it is very unlikely that any race of men or animals could long survive. The scientific conclusion would have been: our theory gives us true results about the movement of the moon and the planets, therefore there is some truth in it, but it does not lead us to foresee the possibility of atoms coming together and being organized into living creatures, *therefore* it is incomplete, *therefore* it overlooks some of the things which atoms actually do.

The root of the matter is perhaps a superstitious feeling that results obtained by looking through a microscope or a telescope are in some mysterious way superior to the knowledge we get in everyday life. We have at times come near to the worship of scientists, to believing that men who work in laboratories can solve all our problems for us. The views of a great scientist on his own science are indeed worthy of respect, for they are based on facts. But by the very act of shutting himself inside a laboratory, a scientist shuts himself out from much of the daily life of human beings. If a scientist realizes this, if he tries to overcome his isolation by paying special attention to current events and by learning the history of mankind, he may be able to apply his scientific training to other departments of life. But if he rushes straight out of his laboratory, full, like any other human being, of prejudice and ignorance, he is likely to make a rare fool of himself.

Euclid's Straight Lines

Beginners in geometry are sometimes puzzled by being told that straight lines have no thickness. We shall never, we are told, meet a straight line in real life, because every real object has a certain thickness. One of Euclid's lines, however, has no thickness. Two lines meet in a point, and a point has no size at all, only position. We shall never meet a point in real life, either, for all real objects have a certain size, as well as a certain position. It is not surprising that pupils wonder how we know anything about objects which no one has ever seen or ever can see.

This difficulty is a good example of the confusion which can come from misunderstanding the methods of abstract thought. We have seen that Euclid's geometry grew out of the methods used for building, surveying and other work in ancient Egypt. This work was done with actual ropes or strings, real 'linen threads' with actual thickness. What does Euclid mean when he says that a line has no thickness, although he is using results suggested by the use of thick ropes? He means that in laying out a football field or in building a house you are *not interested in the size or the shape of the knots* made where one rope is joined to another. If you allowed for the fact that ropes possess a definite thickness, if you carefully described all the knots used by a bricklayer, you would make the subject extremely complicated, and no advantage would be gained. Euclid therefore says, if an actual rope has thickness, *neglect this* in order to keep the subject reasonably simple.

The position is not that Euclid's straight lines represent a perfect ideal which ropes and strings strive in vain to copy. It is the other way round. Euclid's straight lines represent a rough, simplified account of the complicated way in which actual ropes behave. For some purposes, this rough idea is sufficient. But for other purposes – such as teaching Boy Scouts how to tie knots – it is essential to remember that rope has thickness: for such purposes you will obtain wrong results if you think of ropes as Euclidean straight lines.

Similar illustrations might be drawn from any science. Scientific laws are true, within certain limits, for certain types of object. For other types the truth becomes doubtful, and after a certain point, positively misleading.

Some books on mechanics give a very good explanation of the word *particle*. They say: an object is spoken of as a particle when its size is small compared to the distances in which we are interested. The earth sweeps round the sun at a distance of about 90,000,000 miles. Compared with this, the diameter of the earth, say 8,000 miles, does not amount to much, and we might speak of the earth as if it were a point. On a map of the world London might be regarded as a point. According to the atlas, London is $51\frac{1}{2}$ degrees north of the equator. It is not necessary to say whether this refers to Hampstead or the Isle of Dogs.

Readers will find some statements in this book very puzzling if they treat every sentence as being true (so to speak) to ten places of decimals. Particularly within the limits of a small booklet, it is impossible to hedge every sentence round with the remarks and cautions that would be necessary in a text-book. In any case, I do not wish to present people with ready-made opinions. It is up to the reader to approach this book in a mood of sturdy commonsense, with a readiness to criticize and reject anything which, through stupidity, carelessness or lack of space, or through the essential difficulty of saying anything with 100% truth, is misleading. My aim has been to convey a general impression, sufficient to show how mathematicians think.

The Mathematicians on the Second Floor

At this stage there will be a protest from the pure mathematicians, who will say that engineers and other practical mathematicians may think in this rough instinctive way, but that my statements neglect the very important body of people who work on mathematics itself, and neither know nor care what practical applications their work may have.

It is true that pure mathematicians, working in this way under the inner compulsion of an artistic urge, have not only enriched mathematics with many interesting discoveries, but have also created methods of the utmost value for practical men. It is very shortsighted (from the practical point of view) to discourage all work which has no *immediate* practical aim. It pays humanity to encourage the artist, even if the artist does not care in the least about humanity.

What is pure mathematics *about*? It does not seem to deal with any definite thing, yet there is no doubt about its truth, and its discoveries can be used with confidence for practical tasks.

Perhaps the square root of minus one will serve as an illustration. How do we come to study such an unearthly idea? The history of this idea is briefly sketched at the beginning of Chapter 15. Mathematicians first used the sign $\sqrt{-1}$, without in the least knowing what it could mean, *because it shortened work and led to correct results*. People naturally tried to find out *why* this happened and what $\sqrt{-1}$, really meant. After two hundred years they succeeded.

This suggests that pure mathematics first appears as *the study of methods*. Pure mathematicians do not appear on the scene until late in human history: they represent a high level of civilization. The first comers are the practical men, who study the world at first hand, and discover methods which work in practice. Pure mathematicians do not study the natural world. They sit, as it were, upstairs in the library, and study the writings of the practical men. Sometimes the practical men get taken in by a method which usually gives the correct result, but not always (see Chapter 14). The job of the pure mathematicians is then to sort out the methods which are logical (that is, which give correct results) from those which are not.

Pure mathematicians are in touch with the real world, but at second hand. They do not sit by themselves and think. The material they study consists of the books in the libraries of the world. These books do not consist solely of the writings of engineers. The chain is often very long. An engineer consults an 'applied mathematician' (one who studies the applications of mathematics to everyday problems): the applied mathematician consults a pure mathematician: the pure mathematician writes a paper on the question: another pure mathematician points out that the question could be solved if only we knew the solution to some more general problem, and so it goes on. A great literature arises, showing the connexion between different problems. A subject becomes so large that it is impossible to remember all that has been written about it: it becomes an urgent necessity to boil all the various results down into a few general rules. After a century or two, problems are being discussed which seem to have no connexion with the worries of the original engineer. But the connexion is there, even if it is not easy to see.

Is pure mathematics, then, merely the study of how mathematicians think? It certainly is not. Pure mathematicians take very little account of how people actually think. If all the applied mathematicians in the world suddenly went mad, pure mathematics would remain unchanged. Pure mathematics is the study of how people *ought* to think in order to get the correct results. It takes no account of human weaknesses. It would perhaps be more true to say that pure mathematics is the study of how we should have to construct calculating machines, if we decided to do without human mathematicians altogether.

Pure mathematics appeals to those who, like Rupert Brooke, appreciate

> 'the keen
> Unpassioned beauty of a great machine,'

but this is a taste that comes late both in the history of the human race and in the life of most individuals. For the purpose of teaching, it is essential to master the primitive methods of practical mathematicians before attempting to introduce the strict methods of the pure mathematician. The emphasis in this book is laid on practical mathematics, not because practical mathematicians can claim any superiority over pure mathematicians but simply because teaching experience shows that it is necessary to do so.

Nor would I claim any infallibility for the view I have suggested as to how it has come about that men can reason. I have had less time than I would have wished to study the history of mathematics, and of mankind generally. These views I merely believe to be on the right lines. But that most human beings think, and need to be taught, as this theory would lead one to expect – this, from direct experience, I know to be true.

Practical Conclusions

To sum up – successful reasoning is possible only when we have a clear picture in our minds of what we are studying. Imagination is developed, and is made reliable, through practical contact with the real world. Mathematics is difficult when it is presented as something quite apart from everyday life. Mathematical reasoning can grow gradually and naturally, through practical work with real objects. This holds both for elementary and 'higher' mathematics. Only for the 'highest' pure mathematics is the connexion with daily life rather indirect.

CHAPTER 4

THE STRATEGY AND TACTICS OF STUDY

'I have taught mathematics and applied science or engineering to almost every kind of boy and man ... In my experience there is hardly any man who may not become a discoverer, an advancer of knowledge, and the earlier the age at which you give him chances of exercising his individuality, the better.' – John Perry, 1901.

THE two main conditions for success in any sort of work are interest and confidence. People usually pay little attention to these two factors, because they feel (quite rightly) that they cannot make themselves confident or interested by an effort of will.

It is quite true that you cannot increase confidence by an act of will. But neither can you increase the size of your muscles or make your heart beat more vigorously by sitting in a chair and willing it. This does not mean that it is impossible to change your muscular strength or the rate of your heart-beat. If you skip for half an hour you will do both.

Confidence and interest can also be changed *by taking the proper measures.*

The proper measures do not consist in rushing at work like a bull at a gate. It is well known that the effect of too intensive physical training is to destroy the body, not to build it up. The same is true of the mind.

In physical training, some of the vital organs lie beyond the control of our conscious minds. We cannot send direct orders to our heart, our liver, our glands. We have to find exercises depending on the movement of our limbs, on the efforts of muscles which we can control, that will produce the desired effect on the other organs. After a few months of proper training, we do not know what changes have taken place in our bodies, but we feel the benefit and know that changes must have occurred.

In mental training also the decisive changes take place outside consciousness. The test of any system of coaching is not whether it turns out students capable of doing certain tricks, like performing dogs. Such a method is futile and fundamentally degrading. It merely enables people to pass examinations in subjects which they do not understand, and to qualify for posts in which they will be unhappy and inefficient. The real test of any teaching method lies far deeper. With the correct approach, a student finds his whole feeling about the subject changing. He begins to understand what the subject is about, he feels confident that he can master it, he begins to take pleasure in it and to think about it outside working hours. Only when such an attitude has been created does the mind really *grasp* the subject. People show a greater degree of intelligence and knowledge in connexion with their hobbies than in any other department of life.

Lack of Interest

Is it possible to transfer the kind of interest we feel for a hobby, and to use it for the purpose of work? It depends on the reason for your lack of interest.

There are people whose interest is concentrated on one subject. If you feel that you have one single purpose in life, whether it be to paint pictures or to find a cure for cancer – if you feel that this one thing alone matters for you, that everything else – comfort, wealth, respectability, safety, family ties or social obligations – are without significance compared to it – then obviously you have no doubt what you are to do.

Only a few people are thus clear cut in their aims. Most men and women are prepared to fit in, more or less, with the customs they find around them, to work at any job by which they can earn a reasonable living.

There are probably some who fall between two stools – people who could be happy and efficient in some particular type of life, but who lack the self-knowledge or the courage or the determination needed to break away from the life which other people expect them to lead. The war has produced many cases in which people, who had previously been making a rather half-hearted effort to qualify for learned professions, found themselves doing practical work, putting out fires, driving lorries and so on. It was obvious that they had found the type of work for which nature designed them. In a perfect world they would be encouraged to do such work, without a war being necessary. For such people the question is not how to learn mathematics, but how to drop mathematics at the first possible moment.

This then is your first question: to which type do you belong? Are you a person with such a keen interest in some special type of activity, that you can afford to drop other subjects (including mathematics) and succeed as a specialist expert? Or do you belong to the more usual type, that is ready to tackle whatever comes along?

You must decide definitely one way or the other. Either your interests are so far from mathematics that you could never have any use for or amusement from mathematics, or there is something which you accept as worth doing, for which mathematical knowledge is necessary. In answering this question you must make allowance for the fact already mentioned – that the educational system seems specially designed to take all life and interest out of the subjects taught. By mathematics is meant the living subject, not what is taught in many schools.

In some cases, then, the lack of interest goes right down to the roots of personality. But the vast majority of people who hate mathematics do not come under this heading. By far the commonest cause of dislike is the way mathematics has been presented. You can test this for yourself. Do you like puzzles? Do you listen to the Brains Trust, or do crosswords? Do you play bridge, or chess, or draughts? Do you take part in the heated arguments one sometimes hears, such as the question what would happen if passengers in a motor-car threw a cricket ball straight up into the air – would it fall into the car again? Do you take an interest in any sort of scientific or mechanical development, such as radar or aeroplane working? If so, your basic interests are not very different from those of the mathematician. I know a family (by no means high-brow) that was divided into warring factions one Christmas by the car and cricket-ball question. At school, it was the most normal boys who felt most strongly about their own solutions to such problems. This interest in what would happen is close to the interest a scientist feels, and science soon leads to mathematics.

The Removal of Fear

Probably most people would be interested in mathematics, as most people would be interested in music, if they were not afraid of it. Interest and confidence are closely connected. If you find that you can do something, you are pleased. You like the feeling of having mastered nature and the feeling that other people will admire you. You want to do some more of it, and the more you do, the better you become. On the other hand, if you start off with a defeat, the effect is the opposite. Nobody likes to appear a fool. You avoid the subject, or try to make out that you do not bother about it. You decide that you never will be any good, so why waste energy? In any case, you convince yourself, it is no use. All of this has nothing to do with the facts of the case: it is the desperate attempt of a human mind to keep its balance and its self-respect. You probably concentrate on some other subject, or play hard at some game, and say to yourself, 'Well, I may not be able to do algebra, but I am pretty hot at cricket and chemistry.'

In some schools, the excellent custom is followed, when a boy is a complete failure at lessons, to put him on to some useful activity such as carpentry or ploughing. He then becomes sure that he can do *something* well, and he no longer needs to deceive himself about his lessons. He can take the risk of really trying to succeed, since his self-confidence will not be destroyed should he fail.

It is essential, if you are trying to overcome your dread of a subject, to realize what is your first objective. Your first job is *not* to learn any particular result. It is to get rid of fear. You must go back a certain way, and start with work which you are absolutely sure you can do. In learning a foreign language, for instance, it is helpful to get a book written in that language for children just learning to read. However badly you have been taught, you will almost certainly be able to read it. This is your first victory – you have read a book genuinely written for the use of someone speaking a foreign language.

In mathematics it is even more important to go back to an early stage. It is impossible to understand algebra if you have not mastered arithmetic: it is impossible to understand calculus if you have not mastered algebra. If you attempt the impossible, without realizing what you are doing, your morale will suffer.

Apart from this logical necessity, there is also a psychological reason. The chances are that you are still carrying around with you all the feelings of uncertainty that have troubled you during all the different stages of your education. You are still *feeling* the setbacks that you had when you were eight or nine. This feeling will immediately disappear if you go right back to the beginning, and read again the text-books that you then had. You will often find that the difficulties have vanished without your realizing it.

It is for this reason that there are chapters in this book dealing with such things as the multiplication table. You will read these chapters without difficulty. At some stage of the book you will find yourself again puzzled. This means that you have reached the stage where your knowledge of the subject begins to show gaps – at this point, or at some earlier point, your revision must begin. It is nothing unusual to be puzzled by the things you have just learnt. If you keep revising, and are perfectly clear about everything which you have done more than a year or more than six months ago, you need not worry.

A good way to revise is to take a text-book, and look through the examples in it. If you can do them easily, you need not read the book. The examples on some chapters may give you difficulty. If the text-book is one which you first read several years ago, you will probably know whether the results of these particular chapters are much used in later work. If so, you have found out the source of your difficulty in the later work. If they are not important, you may leave them for the time being.

In mathematics it is often necessary to work backwards. If you find a difficulty on page 157 of some book, try to find out why. See if page 157 uses the results of other, earlier pages in the book, or if it uses some fact explained in an earlier text-book. If page 157 depends on pages 9, 32, and 128, read these pages again and make sure that you understand them. If you do not understand them, you cannot possibly understand page 157.

If you still have difficulty, ask someone else to explain the page to you. Notice very carefully if he uses any word, any sign, or any method which is strange to you. If so, ask where this word, sign, or method is explained.

If you can find out what your difficulty is, you are half-way to overcoming it. People often go about with a fog of small difficulties in their heads: they are not quite sure what the words mean, they are not quite sure what has gone before, they are not quite sure what is the object of the work. All these difficulties can be dealt with easily, if they are taken one at a time. Provided the book is written in reasonably simple language, a few minutes with a dictionary should clear up that difficulty.* The next thing is to find out what knowledge you are expected to have before you attempt to understand the proof of a new result. It is possible to make a diagram showing how a book hangs together, how one section depends on previous sections. One should learn a book both backwards and forwards: one should know that the result on page 50 is proved by the result on page 29, and that it is used to prove the result on page 144. (Of course no sane person will learn the numbers of the actual pages on which results occur. But it may be worth while to write in the margin of page 50, 'See p. 29; used, p. 144') Many people learn separate results, but never link them together in this way.

In this book it has not been possible, in every single sentence, to give references to all the remarks, made earlier in the book, that may help towards understanding. If you cannot understand some sentence, underline it. The chances are that somewhere earlier in the chapter, or in the book, a remark has been made that was especially intended to prepare for the difficult sentence. At the first reading you may not have noticed this remark at all. It seemed pointless. Look back for such remarks. If you succeed in finding them, put a note in the margin, 'This explains sentence underlined on page ...'

It may seem to you that this advice does not amount to much, that it is obvious. It may be obvious – but people need a lot of persuading before they do it. As a rule, someone who has difficulty with calculus or trigonometry is not prepared to believe that the real trouble is ignorance of algebra or arithmetic. There is always an examination coming in six weeks, or a year, or whatever it is, and this examination is on calculus or trigonometry – not on algebra and arithmetic. Trying to learn higher mathematics without a firm grasp of the earlier part is like trying to invent an aeroplane without knowing anything about motor-car engines. Until the motor industry had been developed, all attempts at aeroplanes were complete failures.

*I have tried to keep words in this book as short as possible. One or two words may not be known to everyone. It is only fair that readers should take the trouble to look these up.

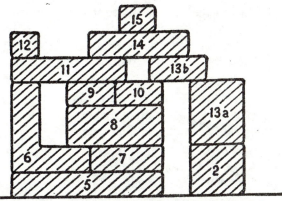

COMMON SENSE & EVERYDAY EXPERIENCE

THE GENERAL PLAN OF THIS BOOK

In this diagram each block represents a chapter. Chapters 1, 3, and 4 are of a general nature, and are not included in the diagram.

Each block depends upon the blocks below. Thus, it is impossible to understand Chapter 11 without first having read Chapters 6, 9, and 10, and Chapters 9 and 10 in turn cannot be understood without Chapter 8, etc.

In some cases, the upper block depends only on a small part of the lower one. For instance, Chapter 8 can be understood without understanding the whole of Chapter 6. In fact, it is only the part of Chapter 6 explaining the meaning of the signs 4^3, 10^5, etc., that is needed for Chapter 8. It is not possible to show this on the diagram.

Chapter 13 is split into two parts. 13a represents the greater part of the chapter, which is quite elementary. 13b represents the end of the chapter, which is more advanced.

If a reader finds difficulty, say in Chapter 10, he may find it worth while to leave Chapters 10, 11, and 12, for the time being, and to read the easier part of Chapter 13.

To revise elementary mathematics takes much less time than people imagine. How many text-books has a student of eighteen used? One on arithmetic, one on algebra, several perhaps on geometry, an elementary trigonometry, perhaps a book on calculus. Geometry we may leave on one side for the moment. How long does it take to look through an arithmetic book and one on algebra, and find out if there is any important result which you missed at school? How long does it take to write down on a sheet of paper a list of the contents of these books, and to put a tick against the results which you thoroughly understand? Not very long. The advantage of doing this is that you begin to see how much (or how little) you have to learn. One tends to think of algebra as a vast jungle of confusion, in the midst of which one wanders without map or compass. It is much better to think of algebra (or that part of algebra which you need to know) as being half a dozen methods, and twenty or so results, of which you probably already know 60%. Nor need you revise the whole of this at once. Suppose for instance you are finding difficulty with calculus because you do not properly know the Binomial Theorem. Get down your book of algebra, and look up *Binomial Theorem*. Never mind about the proof for the moment. First get quite clear what the Binomial Theorem is. It is full of signs such

as ^nC_r or $\binom{n}{r}$ – different signs are used in different books. These

signs are explained in the chapter on Permutations and Combinations. Again, do not bother about proof. See what these signs mean. Work out a few examples – $^4C_1$ $^4C_2$ and $^4C_3$, for instance. Work these right out, as numbers. Come back to the Binomial Theorem, and take particular examples of it. Put $n = 4$, for instance.* The binomial theorem deals with the expression $(x + a)^n$. Put $x = 10$ and $a = 1$. Work out 11^2, 11^3, 11^4. What is the connexion between 11^4 and the numbers worked out above? Work out 101×101 and $101 \times 101 \times 101$. What do you notice about 11×11 and 101×101? What do you notice about $11 \times 11 \times 11$ and $101 \times 101 \times 101$? The same numbers turn up both times? Do you think the same numbers will turn up in 1001×1001 as in 11×11? In $1001 \times 1001 \times 1001$ as in $11 \times 11 \times 11$? If so, you are not far from discovering the Binomial Theorem for yourself. (If you are not clear what 11^4 stands for, the remarks above will be meaningless to you. What 11^4 means is explained in Chapter 6.)

*If you are in the fortunate position of having never been taught algebra, and therefore having no mistaken ideas about it, take no notice of this paragraph. The meaning of algebraic signs is explained in Chapter 7.

In this way, tracing back and back, you get to know the parts of algebra which are useful for calculus. You know at least what the Binomial Theorem is, and how it helps you to write down $(1001)^4$, even if you cannot prove it. When a book or a lecturer refers to the Binomial Theorem, you will be able to follow the use that is made of it. When you are thoroughly familiar with the usefulness and the meaning of the Binomial Theorem, it *may* be worth your while to study the proof. (Some books contain very dull proofs. Look for a book with a proof that is short and that appeals to you.)

Reading with a Purpose

We have just been using an algebra book in a special way – with a purpose. We have not tried to read the whole book. We have taken no notice of any chapters, except those which are necessary for an understanding of the Binomial Theorem. You may not think this is much of a purpose, but it is better than none at all. You will be surprised how much more sensible a text-book becomes if you use it in this way. You have a definite interest in getting this information – it will save you from getting any further behind with your work. You are not littering your mind with all the information in the book. You are learning only things which you need to learn.

All mathematics grew rather in this way. Someone wanted to do or make something: it was impossible to do it without mathematics: so mathematics was studied, and the *purpose* gave meaning and unity to the work done. A very simple example – try to make a model mansion with gables and attics in the roof, by cutting out pieces of cardboard or paper and sticking them together. You will find it is not so easy as it looks to draw the shapes that will be required. From such a problem, scientifically investigated, can arise geometry and spherical trigonometry. By working and experimenting with this problem of the toy-maker and the architect, you will unconsciously acquire the type of imagination necessary for studying geometry, trigonometry and solid geometry.

Interest is a peculiar thing. There are hundreds of things in which you feel you *ought* to be interested – but for which (to be honest for once) you do not give a hang. There are hundreds of other things – odd remarks, pointless little stories, tricks with matches, stray pieces of information – which seem to have no use in life, but which stay in your memory for years. At school we read a history book by Warner and Marten. No one remembered the history (this was no fault of the authors). But there were certain footnotes in it: one about a curate who grew crops in the churchyard and said it would be turnips next year; a lady who blacked out a picture and said, 'She is blacker within'; a verse about someone longing to be at 'em and waiting for the Earl of Chatham – everyone knew these years after they left school. These were the things that really interested us.

If you want to remember a subject and enjoy it, you must somehow find a way of linking it up with something in which you are *really* interested. It is very unlikely that you will find much entertainment in text-books. If you read only the text-books, you will find the subject dull. *Text-books are written for people who already possess a strong desire to study mathematics: they are not written to create such a desire.* Do not begin by reading the subject: begin by reading *round* the subject – books about real life, which somehow bring in the subject, which show how the subject came to be needed.

In any reasonably large town, a public library gives an easy way of finding good books. Nearly all libraries use the same method of indexing books, the Dewey Decimal System. Take a look at the books between 501 and 531. On the open shelves of Manchester Central Reference Library, inside an hour and a half, I found the following books, and glanced through their contents. I give them just as I jotted them down – pass rapidly over any book which does not appeal to you. The reference numbers are given.

510.8 Horsburgh: *Modern Instruments of Calculation*. Do not try to read this straight through. Photographs of calculating machines, round about page 26. If you hate arithmetic, why not make a calculating machine for yourself?

510.2 Mellor. *Higher Mathematics for Students of Chemistry or Physics.* An excellent book, but do not try to read it until you are ready for it.

515. Abbott. *Practical Geometry and Engineering Graphics.* A book full of illustrations. It covers problems rather like that of the model house. How do you cut a flat sheet of metal to make a stove-pipe with a bend in it? What curve is best for making gear wheels?. Glance through the whole book, see what subjects interest you, then trace backwards, and see what type of mathematics is needed for each. Fairly technical language. Beginners should rest content with a general impression.

523. Serviss. *Pleasures of the Telescope.* Very simple indeed. Well illustrated. Contains star maps. Will appeal to artistic people. Useful for airmen and sailors, who may steer by the stars in emergencies.

522.2 Bell. *The Telescope.* Intended for the makers of telescopes and field-glasses. Only a small part of the book consists of mathematics. Best method – read through the book: note anything you cannot understand; then consult an elementary book on Optics (under the number 535). Try your hand at designing a telescope, a microscope, a magic lantern or cinema projector, an epidiascope, a camera obscura. The advantage of doing your own design is that you can use any 'scrap' – old spectacles, magnifying glasses, etc. – that you may have. Very simple geometry is sufficient for this purpose, if you find the right method.

526.8 Hinks. *Maps and Survey.* Chapter 2 explains why it is necessary to have maps. Chapter 8 deals with the kind of map made by an explorer, and Chapter 10 with the rough survey made by the first settlers in a frontier town. Chapter 12 shows how maps are made from aerial photography. By reading *the right parts* of this book, a beginner in trigonometry can get a useful background – how to make a rough plan of a field, etc. The book also brings out some unexpected connexions between practical life and scientific questions: the exact shape of the earth and observations of stars are needed to make a map of a big country like Africa; it is hard to make a proper map of India, because the Himalayas are heavy enough to exert a noticeable pull on a plumb-line, and cause it not to point straight at the centre of the earth.

While on the subject of maps, *A Key to Maps*, by Brigadier H. St J. L. Winterbotham, may be mentioned. Among other things, it tells hikers how to see from a map what the view from any place will be like. Many libraries have, or can obtain, this book.

530.2 Saunders. *A Survey of Physics*. In the words of the author, 'The reader will be introduced to some of the mysteries of nature, as well as to many ingenious inventions of mankind.'

531. Goodman. *Mechanics Applied to Engineering*. Contains a great amount of information. I am not sure how it will appear to beginners. As with all other books, look through it, learn anything you can, but do not be distressed if there are parts of the book you cannot follow at all.

You may find something to interest you under 385, Railways; 620.9, History of Engineering; 626, Canals. If you are particularly interested in any subject, the library assistants will tell you where to look. Look right through the catalogue, under any section that interests you. It pays to spend a long time searching for an interesting book on the subject, rather than to read half a dozen books that will bore you.

It is often good policy to read a book of which nine-tenths merely reminds you of things you already knew, while the remaining one-tenth is new. Your mind will then have plenty of energy to learn the new facts. Do not make a great effort to remember every detail. Anything that interests you will stick in your mind. If you find some useful piece of information that may be needed later, write it down in a note-book kept for the purpose. Your aim should be to have in your mind a general view of the subject, in your desk a collection of exact facts that you can use for any particular problem.

Books on the History and Teaching of Mathematics

If you find these suggestions of any use, if by browsing in a library or by looking about you in the street you hit on anything which you genuinely like and want to know more about (where there's no will, there's no way), you will soon find yourself becoming a specialist in this. It may be anything from radar to how drains are fitted, provided it beckons you on. As you get to know more about this question, you will get impatient with popular introductions, you will find yourself wanting a complete answer to questions, the professional way of dealing with the subject. You will find yourself pulling down bulky volumes that seemed infinitely dry a year ago. You will not read them from cover to cover. You will search with a skilful eye for the paragraph or two that deals with what you want to know at the moment. And you will realize that, while you may not now be interested in other subjects, if you were to become interested in anything, however complicated, you could deal with that too in the same professional way. This confidence, this freedom from fear, is the main thing that distinguishes the expert. An expert does not need to know much. He must know *how* and *where* information can be found.

As the subject you have chosen for your hobby becomes better known to you, you will begin to realize how much like yourself were the men who worked at it and discovered it. When you reach this stage, you may find it useful to have some idea of the dates at which these men lived. There are various reasons for this. (i) By noticing the dates you can get an idea of how much of the subject you know. For instance, if you find that the mathematics you know was all discovered before 1800, you will realize that there is much yet to learn. The nineteenth century saw tremendous mathematical activity. You will not make the mistake of trying to research on questions for yourself without first making some effort to find out if the problem that puzzles you has already been solved. (ii) If you know how much of the subject was known at any time, it is often much easier to see how particular discoveries were suggested by things already known. This helps you to understand the subject. (iii) If you are baffled by something, reading the history of that discovery may help you. The life of the actual discoverer is often very helpful: the attempts he made, the experiments he carried out may supply the clue. In this way you can outflank your difficulty by reading round it – much better than battering your head against it. Far too little use is made of history in teaching mathematics.

In choosing an historical book, as with any other book, look for one which appeals to you, and do not be worried if you cannot read the whole book. Read what you can.

It may also help you to read a book on the teaching of mathematics. There is no ideal way of teaching. What suits one student is useless to another. A teacher who has to deal with a class of fifty has an almost impossible job. If you read a book on teaching, you will find that there are several entirely different ways of tackling the subject. You may feel that you would have done much better if you had been taught by one of these methods. Note the names of the people who developed this method, and see if there are any books by them in your library. *The Teaching of Mathematics*, by J. W. A. Young (1911), contains a description of several movements in mathematics teaching, and the author is human and enlightened. In it will be found a great number of references, on which further reading can be based. One of the reformers mentioned by Young is Professor John Perry, whose *Address to the British Association*, 1901, is the source of the quotation at the head of this chapter. It is well worth reading, both for Perry's speech and for the remarks (mainly approving) of the leading mathematicians of the day. Anything by Perry is worth reading. His book *Calculus for Engineers* may be mentioned. It is forty years since Perry gave this lead. If parents, teachers, and teaching authorities today were fully aware of what was said in 1901, much mental suffering among children would be prevented. The tide is undoubtedly flowing in that direction. It still has a long way to flow.

PART II

ON CERTAIN PARTS OF MATHEMATICS

CHAPTER 5

ARITHMETIC

'One, two, plenty.'
– Tasmanian Method of Counting.

ARITHMETIC plays a very small part in mathematics, especially in the higher mathematics. Geometry, as we have already seen, can be studied direct from diagrams, in which simple numbers – 3, 4, 5, etc. – occasionally occur. The higher one goes, the less arithmetic is likely to be used. This is why there are so many stories about famous mathematicians quarrelling with tram conductors about their change, and being wrong.

Arithmetic does depend on certain things which have to be learnt by heart, such as the multiplication table, addition and subtraction tables. These operations can be carried out by machines, and to a certain extent anyone who learns arithmetic has to become a machine. For instance, a clerk adding up long sums of figures does not need to think deep thoughts about the nature of a number. It is sufficient for him if the sight of 7 and 8 immediately bring the number 15 into his mind.

Whilst it is possible to teach arithmetic in a purely mechanical manner, it is certainly not desirable to do so. Even for the simplest operations, it is easier to remember what has to be done if one knows the reason. For anyone who wants to go on to the other branches of mathematics, mechanical learning is fatal. No one has yet invented a machine that will think for itself. It is a pity that there are still schools (especially girls' schools) where arithmetic is still taught on the lines of 'You do this, then you do that' – as though the subject were some form of religious ritual.

Arithmetic is not a difficult subject to discover for oneself. There are many things which lie on the edge of arithmetic. For instance, when an operation is being carried out in hospital, a nurse has a board with hooks, holding all the things that will go into the patient and must come out again. Before the patient is sewn up, the nurse must see that no hook is empty. This procedure is not counting, but it is very near to it. When we count on our fingers (the original method!) we are simply using fingers instead of hooks.

Counting with fingers (or fingers and toes) will do only for numbers up to ten (or twenty).* Team-work is needed to go further. If a friend is willing to be nudged each time you reach ten, and to count the nudges on his or her fingers, it is possible to reach a hundred. With six people, a million can be counted thus, though the sixth person can go to sleep most of the time. (I do not see why counting by this method should not actually be carried out in classes for young children, in order to explain what is meant by a number such as 243. In playing hide-and-seek children of their own free will count up to quite large numbers, and seem to enjoy it.)

The same idea, in essentials, is used in the devices which measure how far a motor-car or bicycle has gone. Each wheel, on reaching ten, 'nudges' the next. Adding machines are made on similar lines.

The imagination can be further aided if actual objects (say matches) are being counted. The first person ties the matches into bundles of ten. The second person takes ten bundles, and puts them into a box. Ten boxes go into a bag, ten bags into a sack, ten sacks into a truck, ten trucks form a train – the latter stages in imagination only! At the same time, the progress of the work could be exhibited, as on a cricket score-board. Quite soon, the symbol, 127: the sound, 'one hundred and twenty seven': the picture, one box, two bundles, seven matches: would be welded together in the mind of each child.

All the operations, such as adding together 14 and 28, subtracting 17 from 21, dividing 84 into three equal parts, can be

---

* Some entertaining details of primitive methods of counting may be found in E. B. Tylor, *Primitive Culture*, Chapter 7; Tobias Dantzig, *Number, the Language of Science*, Chapters 1, 2.

carried out, first by experiments with the actual objects: secondly with the objects and with 'score-boards' simultaneously: finally, by written work alone.

In the Montessori method, the tables of addition are taught in some such way. The children have sticks, representing the numbers one to nine, and have to arrange these so as to get ten units in each row: thus –

X XᴄᴅXᴄᴅXᴄᴅXᴄᴅX ᴄᴅXᴄᴅX   1 + 9 = 10,

XᴄX XᴄᴅXᴄᴅXᴄᴅXᴄᴅXᴄᴅX   2 + 8 = 10, etc.

It is quite useful to have squared paper, cut into strips, ten squares in breadth. The squares are numbered –

1	2	3	4	5	6	7	8	9	10
11	12	13	14	15	16	17	18	19	20 etc.

The addition table can then be worked out, by pasting on strips of correct length. If two numbers, such as 7 and 6, require more than one complete row, the extra squares are cut off and go to the next row.

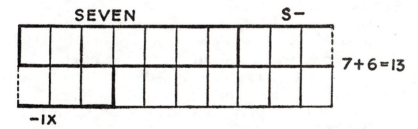

I have heard successful mathematicians say that in adding 7 and 6 the thought was present 'in the back of their minds' that 3 of the 6 units were needed to bring 7 up to 10, so that 3 was left over to provide the odd units.

This method can be extended to the multiplication table, by repeatedly sticking on strips containing the same number of squares. Quite striking patterns emerge. How crude the table for two times

2	4	6	8	10
12	14	16	18	20

is, compared to 6 times! –

					6				
	12						18		
			24						30
					36				
	42						48		
			54						60

The crudest of all (and the easiest to learn) is 10 times: then 5 and 2; then 9 and 3; then 4, 6, 8; the most subtle is 7 – it would make a good wall-paper.

The appearance of pattern appeals to the artistic side, which is strong in children. Good mathematicians are very sensitive to patterns.

Patterns also suggest questions. Why does '3 times' have a pattern rather like that of '9 times'? Why are '5 times' and '2 times' arranged in upright lines?

It was said of Ramanujan that every number seemed to be his personal friend. One should try to present arithmetic to children in such a way that they come to realize the 'personality' which each number possesses.

Owing to the accidental fact that we possess ten fingers, the multiplication tables depend on this number 10. If we had eight or twelve fingers, the patterns would be different.

One can, however, represent multiplication, quite apart from the question of fingers, by means of rectangles. A piece of linoleum, 3 yards by 2, costs six times as much as a piece 1 yard square. We can, if we like, think of 2 times 3 as the area of a rectangle 2 by 3.

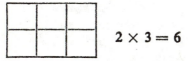

$$2 \times 3 = 6$$

This idea may be helpful in connexion with fractions. We may explain the meaning of $\frac{2}{3} \times \frac{5}{7}$ by saying that it represents the area of a piece of linoleum that measures $\frac{2}{3}$ yard by $\frac{5}{7}$ yard.

Multiplication of fractions seems to cause trouble. Children are often puzzled why two-thirds *of* five-sevenths should be the same as two-thirds *multiplied* by five-sevenths. They see no connexion between 'of' and 'times'. This difficulty is very largely one of language. It is quite natural to say that one field is 3 times or 4 times or $3\frac{7}{8}$ times as large as another. It is perhaps less usual to say a field is $\frac{7}{8}$ times as large as another, more usual to say it is $\frac{7}{8}$ of the size of the other. It is at any rate clear that to draw an area $3\frac{7}{8}$ *times* as large as this page one would take 3 pages, and add this to $\frac{7}{8}$ *of* a page.

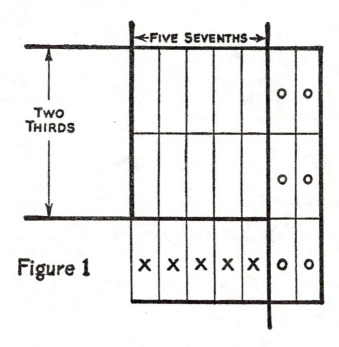

Figure 1

Multiplication of fractions is often taught purely by rule, but it is easy to show why the rule works. Consider for example, $\frac{2}{3}$ of $\frac{5}{7}$. Let us take 1 square yard of linoleum and see what two-thirds of five-sevenths of a square yard looks like. To obtain five-sevenths we must first divide the linoleum into seven equal pieces – by the upright lines in Fig. 1 – and take five of these pieces. If we cut along the heavy upright line, the piece to the left contains five-sevenths. We now require two-thirds *of this*. The level lines divide the whole figure into three equal parts. By cutting along the heavy flat line we shall obtain a piece which is two-thirds of five-sevenths. After the first cut the pieces marked with noughts are removed: after the second cut, those marked with crosses.

This figure shows how to represent $\frac{2}{3}$ times $\frac{5}{7}$ as a single fraction We have divided our square yard into 21 pieces – all the same size and shape. The rectangle, $\frac{2}{3}$ by $\frac{5}{7}$, contains 10 of these little pieces. Each piece is $\frac{1}{21}$ of a square yard, so our answer is $\frac{10}{21}$. In fact, we have found the rule for multiplying fractions –

$$\frac{2}{3} \times \frac{5}{7} = \frac{2 \times 5}{3 \times 7}$$

A common mistake found in examination papers is due to pupils mixing up the rule for adding and for multiplying fractions. They write, for instance –

$$\frac{1}{3} + \frac{3}{5} = \frac{1+3}{3+5}$$

which is complete nonsense, since the answer thus obtained, $\frac{4}{8}$, reduces to $\frac{1}{2}$, which is *less* than $\frac{3}{5}$.

With parrot learning, such a mistake is quite natural. '$\times$' has been turned into '$+$': that is all. Such a mistake is much less likely in a pupil who has made experiments with $\times$ and $+$, and has come to *feel* the entirely different meanings of these two signs.

The reader may care to work out a diagram which illustrates the correct way of adding $\frac{1}{3}$ and $\frac{3}{5}$.

Decimals

No difficulty at all should be found in teaching or learning decimals. Decimals can be demonstrated by exactly the same 'team work' as was suggested for whole numbers.

The measurement of a line is a convenient illustration. A metre is a French measure, not much different from a yard. A decimetre is one-tenth of a metre; a centimetre one-tenth of a decimetre; a millimetre one-tenth of a centimetre. A line whose length is 1 metre, 3 decimetres, 2 centimetres, and 5 millimetres is written for short as 1·325 metres.

While in English measure it is not simple to turn 2 yards 1 foot and 3 inches into inches, it is at once obvious in French measure that 1·325 metres is 1325 millimetres, or 132·5 centimetres, or 13·25 decimetres.

An ordinary school ruler has millimetres, centimetres, and decimetres marked on it. It is therefore quite easy to build up the length mentioned above – one strip a metre long, three strips of a decimetre, two of a centimetre, five of a millimetre.

Addition of decimals is the same as addition of whole numbers. Multiplication of decimals can be illustrated by means of rectangles, just as was done with ordinary fractions.

Negative Numbers

A picture in *Punch* during the 1914–18 war showed an official saying to a farmer, 'My dear sir, you cannot kill a whole sheep at once!'

This absurd remark illustrates the fact that *fractions* have no meaning for certain things: you cannot have half a live sheep: you cannot tear a sheet of paper into 3½ pieces. But fractions have a meaning in other connexions: it is quite easy to have 3½ feet of lead piping.

In the same way, there are times when you cannot speak of numbers less than 0: there are other times when you can.

A man may have no children, but he cannot have less than none. A box may have nothing in it: it cannot have less than nothing.

But there are examples where we can go below 0. For instance, in the Fahrenheit system of temperatures, water freezes at 32 degrees, a mixture of water and salt freezes at 0 degrees, and it is possible to have temperatures much colder than this. These temperatures are written with a *minus* sign. Thus —10 degrees means the temperature 10 degrees colder than 0 degrees. A temperature of —22 degrees is met with in refrigerators using ammonia. Note that —22 degrees is *colder* than —10 degrees.

In the same way, we may deal with heights and depths. If a bomb falls into the sea from a height of 50 feet we can trace its descent from 50 feet, to 40, 30, 20, 10, and 0 feet above sea-level. But the bomb need not stop at sea-level. It may descend 10 feet below sea-level, and we may speak of this as a *height of* —10 *feet*.

A man who is in debt to the extent of £1 is worse off than a man who has no money and no debts (like a tramp). The tramp at least is free. If we call the tramp's fortune £0, we may call that of the other man £(—1). When you own £(—1), someone has to give you £1 before you reach the position of owning nothing. To own £(—100) means to be bankrupt to the extent of £100. Again —100 is *worse* than —1. If a minus sign is in front of a number, the order of the numbers is turned right round. £(—1) represents a *better* fortune than £(-10,000).

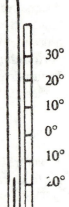

30°

20°

10°

0°

10°

20°

In the same way, an army retreating at 10 miles an hour might be spoken of as 'advancing at —10 miles an hour'. If the army is moving at '—1 miles an hour', that is *better* than moving at '—10 miles an hour'.

A minus sign turns everything upside down, like the reflection of trees and houses in a river.

For a long time, mathematicians felt that it was unfair to use minus numbers (also called *negative numbers*), but it was found in the course of time that minus numbers could be used, added, subtracted, multiplied and divided, and useful results obtained.

Working With Negative Numbers

We may see how to use minus numbers, if we think of ordinary numbers as meaning something *given*, minus numbers as something *taken away*. We might think of 5, for instance, as a five-pound note, or as something given five times: —5 would then mean a bill for £5, or something taken away five times.

Very often we put brackets round minus numbers. For instance, if we want to say 'add —4 to —3', it looks rather queer if we write simply —4 + —3. So we write (—4) + (—3). This means that the thing in the first bracket, —4, has to be added to the thing in the second bracket, —3. (—4) — (—3) would mean that we had to take —3 away from —4.

What would these things mean in practice? We might say that —4 added to —3 meant that a man already owed £4, and then he got a further bill for £3, so that he would be altogether £7 in debt. Or an army might have lost 4 miles of territory. and then lost another 3 miles. The second *loss* has been *added* to the *first*. In either case, we see that a *loss of* 4 together with a *loss of* 3 is the same as a single *loss of* 7. In the signs of arithmetic, (—4) + (—3) = (—7).

In the same way, if we have to add 4 and —3, this means a *gain* of 4 followed by a *loss* of 3, which is clearly the same as a single gain of 1. In short, 4 + (—3) = 1. In fact, 4 + (—3) means exactly the same as 4 —3.

There is nothing new in this, except the signs, and these signs are often used in ordinary life, to show the changes in trade. in unemployment, in the state of parties at elections, + for increases, — for decreases.

Subtracting minus numbers is sometimes a little confusing at first. It is well first to be clear what subtraction means. 7 —3 = 4 means that a man with £7 as compared with a man having £3 is £4 better off. Subtraction means comparing two things. And we can compare losses as well as gains. An army which has lost 200 men is better off than an army which has lost 1,000 men to the extent of 800 men's lives saved. A *loss* of 200 is written for short —200. A *loss* of 1,000 is written —1,000. To *compare* the two we subtract. $(-200) - (-1,000) = 800$. Note that there is no *minus* sign with the 800. If two opposing armies begin by being equal, the one that loses 200 men is *stronger* than its opponent, who loses 1,000, by 800 living men.

We could instead interpret $(-200) - (-1,000) = 800$ as meaning that a man bankrupt for £200 is better off than a man bankrupt for £1,000 by £800. Or we could say that a wreck 200 feet below sea level is higher up than one 1,000 feet down by the amount of 800 feet. It would be correspondingly easier to salvage.

Multiplication? We can only mention this briefly. We may think of 4×5 as meaning, 'Give someone four £5 notes.' This is the same as giving £20, and $4 \times 5 = 20$.

What would $4 \times (-5)$ mean? —5 stands for taking away £5, or for a bill for £5. $4 \times (-5)$ means 'Four bills for £5', the same then as 'A bill for £20'. So $4 \times (-5) = -20$.

$(-4) \times 5$ comes to much the same thing. It would correspond to 'Take away four £5 notes', that is, 'Take away £20.' So $(-4) \times 5 = -20$.

The trickiest case is $(-4) \times (-5)$. Treating —5 as meaning 'A bill for £5' and —4 as 'Take away 4 times', $(-4) \times (-5)$ would mean 'Take away 4 bills for £5'. If the postman comes to you and says, 'I think you have four bills for £5. They should have been delivered to the family next door,' you find yourself £20 better off than you would have been, had the bills really been meant for you. 'Better off' means +. So the effect of two minuses, *multiplied together*, is to give +. We conclude that $(-4) \times (-5) = 20$.

You may very likely feel that this is a tremendous song-and-dance about nothing at all. Everyone knows that you are better off if your creditors destroy your I.O.U.s. Why make all this fuss about + and — signs? The answer is that we are not going to use minus signs simply to find out what happens to people in debt. Rather we are going to be concerned with formulae, such as $y = x^2 - 3x$, or $y = (x - 1)(x - 2)$, in which — signs may occur. That is why we have to know how to handle minus signs. What formulae are, and what uses can be made of them, will appear in later chapters.

Imaginary Numbers, or Operators

You will notice that 3×3 is 9, and -3×-3 is also 9. There is no ordinary number (either $+$ or $-$) which, when multiplied by itself, gives -9. 'Two minuses make a plus.'

It is usual to call 3×3 'the square of 3': 9 is the square of 3. 9 is also the square of -3. 3 and -3 are called the 'square roots' of 9.

Every possible number has two square roots: one $+$, one $-$. The square roots of 4 are $+2$ and -2. The square roots of 10 are (very nearly) $3 \cdot 16$ and $-3 \cdot 16$.

But negative numbers do not seem to have any square roots. -9 has not, nor has -4 or -10. So far as square roots are concerned, negative numbers are the Cinderellas of mathematics. But mathematicians have succeeded in finding a sort of substitute. If they have not found a man for Cinderella, they have at least found a robot. These robots are called Operators. Operators are not Numbers, but they can do many things which real numbers do, just as robots do *some* things that men do. For instance, you can multiply operators. And a particular operator, called '3i', is such that 3i times 3i is -9, while another, called 'i', is such that $i \times i = -1$.

This 'i' sounds like a mathematical fairy story. The interesting thing is that for many very practical purposes – such as wireless or electric lighting – 'i' is very useful indeed. We shall later on explain what 'i' is, and show that there is nothing mysterious about it at all.

EXERCISES

Questions of Pattern

In questions 1–4, squared paper can be used, as was done in connexion with the multiplication tables. For each question, the reader should consider what is the best number of squares to have in each row. For instance, in the first question, if we take 9 squares in each row (as illustrated) what is happening becomes clear.

1. Albert Smith and his wife Betty are both in the forces. Albert is off duty every ninth evening: his wife is off duty every sixth evening. Albert is off duty this evening: Betty is off duty tomorrow evening. When (if ever) will they be off duty the same evening?

A	B						B	
A				B				
A	B						B	
A				B				
A	B						B	
A				B				

In the pattern, each square represents an evening. It is marked A when Albert is free, B when Betty is. It will be seen that the Bs always come in the 2nd, 5th or 8th column – never in the 1st, where the As are. The answer is: They are never free together.

2. In Question 1, would it have made any difference, if Betty had been free every fifth night, instead of every sixth?

3. On a fire-watching rota Alf watches every third night, Bill every fourth night, Charlie every fifth night, Dave every sixth, and Edward every seventh night. All the men begin their duties on the same night, a Friday.

How long will it be before Alf and Bill are again on duty together? Alf and Charlie? Bill and Charlie?

Will Alf and Charlie ever be on duty together without Dave being there?

On Fridays, when not fire-watching, the men attend a club. How often does Alf miss club night? How often do the others miss it?

Is there any night of the week for which Alf can make a regular appointment? Or does he, sooner or later, do duty on every day of the week? How are the others placed in this respect?

(Use squared paper, with 7 in a row, so that all the Sundays come in one column, etc.)

4. Can you see any principle underlying the answers to Question 3? Can you answer the question – How long will it be before Alf, Bill, Charlie, Dave, and Edward are all again on duty the same night? What night of the week will this be?

5. Two men are walking side by side. One takes four steps in the same time that the other takes three. At the beginning they step off together. In what order will the sound of their footsteps be heard?

(Draw a line, representing the passing of time, and mark on it the moments when the feet of the two men strike the ground.)

6. Question 5 may be varied as much as desired – 5 steps against 4, 7 steps against 5, etc. – and the diagrams drawn.

7. A box has to be made, which can be exactly filled, *either* with packages 6 inches long, *or* with packages 8 inches long, placed end to end. What is the smallest length that the box can be made?

Two Questions for Research

A mathematician does not usually just solve a problem and then forget all about it. If he has solved a problem. he starts altering the conditions of the problem. and sees if he can still solve the problem. He wants to make sure that he will be able to answer any question *of that type* that may confront him in future. He wants to discover if there is any simple principle or rule underlying the problem. You can see from the two examples below how a mathematician goes to work.

The Cup-Tie Problem. If 7 teams enter for a knock-out competition. how many matches will have to be played? (It may be assumed that there are no draws or replays.)

In the first round, one team must be given a bye, and 3 matches will be played. This will leave 4 teams for the second round, the semi-final. There will be 2 matches in the semi-final. There is 1 match in the final round. The total number of matches is 3 + 2 + 1. The answer is therefore 6.

The particular problem is easily answered. But suppose, instead of 7 teams, 70 or 700 had entered. How many matches would be necessary then? It would take rather a long time to work out directly. It would help us a lot if we could find a simple rule that saves working out all the rounds separately.

To see if there is any simple rule, a mathematician would start working out the simplest possible cases. If only 1 team entered, no matches at all would be necessary. 2 teams could settle the question by playing 1 match. Work out how many matches would be necessary in the cases where 3, 4, 5, 6, etc., teams entered. You will soon see that there is a simple rule connecting the number of teams with the number of matches.

Lastly, can you see the reason *why* there is this simple rule? How many matches would have to be played if 2,176,893 teams entered a competition?

The Income Problem. There is a well-known puzzle, as follows.

Two clerks are appointed in an office. Smith is to be paid yearly, starting at £105, and rising by £10 each year. Jones is to be paid half-yearly, starting with £50, and rising £5 each half-year. Which has made the better bargain?

Most people are rather surprised when they see how this works out. All one needs to do is to write down what each receives, as follows –

	SMITH	JONES		
		January–June	July–December	Total
	£	£	£	£
1st Year	105	50	55	105
2nd Year	115	60	65	125
3rd Year	125	70	75	145
4th Year	135	80	85	165

One naturally thinks that a rise of £5 every half-year is the same as a rise of £10 every year. But it is not The yearly salary of Jones rises by £20 each year. He has made a much better bargain than Smith.

This question naturally suggests others. A rise of £5 every six months is as good as a rise of £20 a year. What would have happened if the payment had been quarterly? What is a rise of £5 a quarter worth, in terms of yearly salary? Or monthly pay? What is £1 a month rise worth, in yearly pay? What is 1s. a week's rise in weekly pay equal to?

Or the other way round – what rise every six months is the same as a rise of £10 each year? Every quarter? Every month? Every week?

What is the principle involved? Why do things work out this way?

CHAPTER 6

HOW TO FORGET THE MULTIPLICATION TABLE

'My lord, I have undertaken this long journey purposely to see your person, and to know by what engine of wit or ingenuity you came first to think of this most excellent help in astronomy, viz. the logarithms; but, my lord, being by you so found out, I wonder nobody found it out before, when now known it is so easy.'
– Briggs to Napier (From F. Cajori, *History of Mathematics*)

IF you ask an engineer, 'What is 3 times 4?' he does not answer at once. He fishes a contraption known as a slide-rule out of his pocket, fiddles with it for a moment, and then says, 'Oh, about 12'. This may not impress you very much. But if you say to him, 'What is 371 times 422?' he will give you the answer to this, in just about the same time, and without needing to write down any figures.

What is a slide-rule? How is it made? How was it invented? How is it used?

A slide-rule consists of two scales, on each of which can be seen the numbers 1, 2, 3, 4, 5, etc. These numbers are not spaced evenly, like the numbers on a 12-inch ruler. The distance between

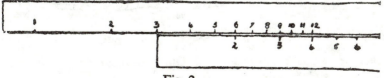

Fig. 2

2 and 3 is less than that between 1 and 2, and the further you go the more closely the numbers are crowded together.

Fig. 2 shows how the engineer would set his slide-rule to find 3 × 4. He pushes the lower scale along, until the 1 on it is opposite the 3 on the upper scale. Now notice how the numbers stand opposite each other. Above 2, there stands 6: above 3, 9: above 4, 12. Above every number on the lower scale, one finds three times that number on the upper scale. So we read off the number above 4, and that gives us our answer 12.

What principle lies behind the working of this instrument? How could anyone have been led to invent it? Why is it possible to make a multiplying machine at all?

We are all familiar with machines which man uses to *multiply* his own strength – pulleys, levers. gears, etc. Suppose you are fire-watching on the roof of a house, and have to lower an injured comrade by means of a rope. It would be natural to pass the rope round some object, such as a post, so that the friction of the rope on the post would assist you in checking the speed of your friend's descent. In breaking-in horses the same idea is used: a rope passes round a post, one end being held by a man, the other fastened to the horse. To get away, the horse would have to pull many times harder than the man.

The effect of such an arrangement depends on the roughness of the rope. Let us suppose that we have a rope and a post which multiply one's strength by ten, when the rope makes one complete turn.

What will be the effect if we have a series of such posts? A pull of 1 lb. at *A* is sufficient to hold 10 lb. at *B*, and this will hold 100 lb. at *C*, or 1,000 lb. at *D* (Fig. 3).

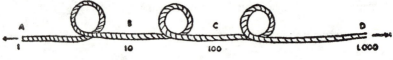

Fig. 3

Each extra post multiplies by 10. One post magnifies by 10: two by 10 × 10: three by 10 × 10 × 10.

As it takes too much space to write long rows of tens, an abbreviation is usually written. 10^2 is written for 10×10. 10^3 for $10 \times 10 \times 10$, and so on. (In the same way, 8^5 would mean $8 \times 8 \times 8 \times 8 \times 8$.)

Thus 10^8 *will represent the effect of* 8 *posts*, and 10^{11} the effect of 11. This is a *multiplying* effect. If we pass a rope round 8 posts and then round a further 11 posts, the effect will be $10^8 \times 10^{11}$. But 8 posts and 11 posts add up to 19, so that this must be exactly the same thing as 10^{19}.

The number of turns required to get any number is called the logarithm of the number. For instance, you need 6 posts to multiply your strength by 1,000,000. So 6 is the logarithm of 1,000,000. In the same way, 4 is the logarithm of 10,000.

So far we have spoken of whole turns. But the same idea would apply to incomplete turns. If you gradually wind a rope round a post, the effect also increases gradually. At first you must bear the entire weight yourself: as the rope winds on to the post, friction comes to your aid, and there will be stages at which you can hold twice, three times, four times the amount of your pull. When one complete turn is on, you will have reached ten times.

Accordingly, $10^{\frac{1}{2}}$ will mean the magnifying effect of half a turn, $10^{2\frac{3}{8}}$ will mean the effect of $2\frac{3}{8}$ turns. And so for any number.

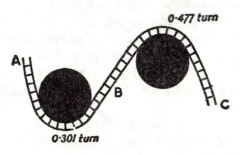

0·477 turn

A

B

C

0·301 turn

Fig. 4

The logarithm of 2 will be that *fraction of a turn* which is necessary to magnify your pull 2 times. This number is usually called 'log 2' for short. Actually, 0·301 of a turn is required to magnify 2 times. $0 \cdot 477$ of a turn magnifies 3 times: so that log $3 = 0 \cdot 477$. (These numbers could be found by experiment.)

We can put this another way round. 2 is the effect of $0 \cdot 301$ of a turn. So $2 = 10^{0 \cdot 301}$. In the same way, $3 = 10^{0 \cdot 477}$.

Now what will happen if we wind $0 \cdot 301$ of a turn on one post, and then $0 \cdot 477$ on the next?

We know that the effect of the first post is to double our effort. If we pull A with a force of 1 lb. it will be sufficient to balance a tension of 2 lb. in the rope at B (Fig. 4). The second post multiplies by 3: 2 lb. at B will balance 6 lb. at C. So 1 lb. at A will hold 6 lb. at C. And, on the two posts together, we have $0 \cdot 301 + 0 \cdot 477 = 0 \cdot 778$ of a turn.

$0 \cdot 778$ of a turn is needed to multiply 6 times. $0 \cdot 778$ is the logarithm of 6. The logarithm of 3×2 has been found by *adding together* log 3 and log 2.

It is not necessary to use separate posts. We can economize in timber by winding the rope again and again round the same post. The only thing that matters is the length of rope *in contact* with the wood. (The post itself must be round. Corners would cause complications.)

If we were given two pieces of rope, and knew that one piece was sufficient to multiply 7 times, and the other sufficient to multiply 8 times, we should only have to join these pieces end to end, to get a piece that would multiply by 7×8.

It is exactly this principle of joining *end to end* that is used in the slide rule. On the slide rule, the distance between 1 and 3 is equal to the length of rope required to multiply 3 times; the distance between 1 and 4 is equal to the length of rope required to multiply 4 times; and in finding 3×4 we place these lengths end to end.

1 of course comes at the end of the scale, as you do not need any rope at all to multiply your strength by 1.

You will now be able to see why numbers crowd together on the slide-rule as we go farther along. 1 corresponds to no rope; 10 to 1 turn; 100 to 2 turns; 1,000 to 3 turns. The distance on the slide-rule from 1 to 10 is the same as that from 10 to 100, or from 100 to 1,000: each of these is equal to 'one complete turn'. But we have only 9 numbers to fit between 1 and 10: there are 90 between 10 and 100, 900 between 100 and 1000. This accounts for the overcrowding of the larger numbers.

If we want to get a set of evenly spaced numbers we have to take a set like 1, 10, 100, 1000 . . . or 1, 2, 4, 8, 16, 32 . . . In the first set, each number is 10 times the previous one: there is 1 turn between each and the next. In the second set, each number is twice the previous one: at each step we add a length of rope equal to 0.301 of a complete turn.

How Logarithms are Calculated

We have explained what a logarithm is, but we have not shown how to calculate it. We have said that, on a slide-rule, the distance between 1 and 7 is the length of rope needed to magnify a pull 7 times – *i.e.*, log 7. But to make an actual slide-rule we should need to know log 2, log 3, log 4, etc., so that we could mark 2, 3, 4 . . . at the corresponding distances.

The only logarithms that have been found so far are those of 10, 100, 1000, etc. We know that these are 1, 2, 3 . . . All this tells us about log 70 is that it must lie somewhere between 1 and 2: for we need more than one turn, but less than 2, to produce any numbers between 10 and 100.

There is one other thing that is vague. We have spoken all along of so many complete 'turns'. But the size of the post has not been specified. We could in fact take a circular post of any size, and pass a rope round it. This arrangement might multiply our pull by less than 10: we could correct this by making the post more rough. If it magnified our pull too much, we could correct this by polishing the post. So that we may suppose 'one turn' to represent any length we like. A slide-rule can be made any size we like. We could, for instance, mark 1 at the end of the scale, and 10 at a distance of one foot. 100 would have to come 2 feet from 1, 1000 3 feet – and by this time we should feel that the whole thing had got quite large enough. Notice that our simple argument has only helped us to mark four points on a yard stick – 1, 10, 100, 1000.

But we could go about the question another way. If we start from 1 and keep *doubling*, we shall also get a set of evenly spaced points: the distance between each point and the next is log 2. (Earlier, we stated that log 2 was 0·301. But *no reason* was given for this statement.) Instead of fixing our scale by taking 10 at 1 foot, suppose we fix it by taking 2 at a convenient distance. We might choose it at an inch from 1. 4, being 2 × 2, must now come at 2 inches; 8, being 2 × 4, will come at 3 inches; 16 at 4 inches; 32 at 5; 64 at 6; 128 at 7; 256 at 8; 512 at 9; 1024 at 10. This slide-rule has turned out to be smaller than the last one: 1000 this time has come just below 10 inches from 1. But that is not the important point. The chief thing to notice is that the first slide-rule had only four points on it – 1, 10, 100, 1000. But our second attempt has given us *eleven points* – 1, 2, 4, 8, 16, 32, 64, 128 256, 512, 1024. This suggests that we shall get still better results by taking, instead of 10 or 2, some number closer to 1, such as 1⅛, or 1·1, or 1·01. It will take more work, with the smaller numbers, to get from 1 to 1000; but once the slide-rule has been made, we shall be able to use it whenever multiplication has to be done, and the work is thus repaid.

Before we leave our slide-rule with eleven points, we may notice that it enables us to get a rough idea of the value of log 2. We saw that 1024 was marked at a distance of 10 inches, so that 10 inches of rope around the post multiply our effort by 1024. But we know that three complete turns multiply by 1000. So that 10 inches must be slightly more than three complete turns. One inch must be slightly more than $\frac{3}{10}$, or 0·3, of a turn. But the figure 2 is marked at a distance of 1 inch. So that 2 corresponds to slightly more than 0·3 of a turn, which is the same as saying that log 2 is just over 0·3. So that our earlier statement that log 2 was 0·301 was at least near to the truth.

How Logarithms were Invented

We made our second slide-rule by a process of continual doubling. We shall now make a better one, using 1·1 instead of 2.

Suppose, then, that we mark on our scale two points, to

represent 1 and 1·1. The distance between them could be, say, $\frac{1}{10}$ inch. We then know that moving $\frac{1}{10}$ inch down the scale (if you prefer, adding $\frac{1}{10}$ inch to the length of the rope) represents multiplication by 1·1. We shall thus be able to mark the points 1, 1·1, 1·21, 1·331, etc., each number being one and one-tenth times the previous one.

This set of numbers is exactly the set that you get if you allow £1 to accumulate at compound interest of 10% over a number of years. Each year that passes increases the sum invested by one-tenth – that is, every year multiplies the amount by 1·1. It was probably the study of tables of compound interest that originally suggested the idea of logarithms to their inventor, Napier.

Some of the numbers found in this way, and the distances at which they have to be marked, are shown in the following table.

Number	Distance (inches)
1·948	0·7
2·143	0·8
2·852	1·1
3·137	1·2
5·053	1·7
6·725	2·0
7·397	2·1
9·846	2·4
10·831	2·5

This shows us that the figure 2 has to be marked somewhere between 0·7 and 0·8 inch; 5 just below 1·7 inches; 7 somewhere between 2·0 and 2·1 inches; 10 a little above 2·4 inches. So 'one turn' corresponds to a little more than 2·4 inches.

This information is still not sufficient for making a really good slide-rule. For instance, we cannot find an accurate position for the figure 7. Our table contains no number between 6·725 and 7·397. We can only guess where 7 lies, between these two numbers. Our slide-rule would, in fact, be liable to errors of about 10%: owing to the fact that the numbers are obtained by adding on 10% at a time – we cannot expect any higher degree of accuracy.

We can use this table to find the logarithms of the numbers 2, 3, 4, etc., but the results for this also are likely to be crude. 'One turn' is the distance corresponding to 10: we guess it to be 2·42 inches. To 2 corresponds a distance between 0·7 and 0·8 inch – perhaps 0·73. If we express 0·73 as a fraction of a 'turn', we shall have an estimate of log 2. 0·73 divided by 2·42 gives 0·3016 – a suspiciously good result for a guess!

The reader will easily see how accurate slide-rules and tables of logarithms can be made, by using a number such as 1·000001 for repeated multiplication. Napier, in making the first tables of logarithms, used 1·0000001.

It is not, of course, necessary for us to make our own tables of logarithms. This work has been done once and for all. The only advantage to be gained from making your own table of logarithms and your own slide-rule is the insight this gives into the underlying principles.

The method of making logarithm tables, described above, shows clearly why such tables can be used for multiplication. We found, for instance, that multiplying by 1·1 seven times was the same as multiplying once by 1·948; while multiplying seventeen times by 1·1 was the same as multiplying once by 5·053 (see the table above). So 1·948 × 5·053 corresponds to 7 multiplications by 1·1, followed by 17 more – that is, it corresponds to 24 multiplications. And this (from our table) corresponds to 9·846. So 1·948 × 5·053 = 9·846.

The method is clear. 1·948 is the 7th number; 5·053 is the 17th; 7 + 17 = 24; the 24th number in the table gives the answer.

In making our slide-rule, we put 1·1 a tenth of an inch from 1, 1·948 seven times as far away, and so on. It does not matter how far 1·1 is from 1, so long as 1·948 is seven times that distance, 5·053 seventeen times as far, etc. The argument will still hold.

In ordinary logarithm tables, log 10 is 1. We saw, on our slide-rule, that 10 came between 24 and 25 times the distance of 1·1. If we choose a distance for 1·1 which lies somewhere between $\frac{1}{24}$ and $\frac{1}{25}$ inch, we shall get 10 coming at a distance of 1 inch. The distance corresponding to any number will be its logarithm.

This change of scale rather camouflages the simple relation,

$7 + 17 = 24$. In the logarithm tables, $1 \cdot 1$ corresponds to $\cdot 0414$; $1 \cdot 948$ to $\cdot 2896$; $5 \cdot 053$ to $\cdot 7036$. In the surface, there is nothing simple about these numbers. But notice these facts. (i) log $1 \cdot 1$ lies between $\frac{1}{24}$ and $\frac{1}{25}$, as we expected, (ii) log $1 \cdot 948$ is seven times log $1 \cdot 1$, and log $5 \cdot 053$ is seventeen times log $1 \cdot 1$. The simple relations are still there. The change of scale does not in any way alter the method: to multiply numbers we add their logarithms.

If we are asked to calculate an expression such as 12^{35} – i.e., the effect of multiplying by 12 thirty-five times – this is easily done. To multiply by 12, one has to add log 12 to the logarithm of the number being multiplied. If one multiplies by 12 thirty-five times, one will thus add log 12 thirty-five times; log 12 is $1 \cdot 0792$. Thirty-five times this is $37 \cdot 772$. $37 \cdot 772$ is the logarithm of 5916 followed by 34 noughts! So this, roughly, is the effect of multiplying by 12 thirty-five times. It would take rather long to find this result by any other method.

A Musical Slide-Rule

One well-known object is in effect a slide-rule – a piano keyboard. The strings at the bottom end of a piano vibrate slowly: as one goes up the keyboard, the rate increases. An octave corresponds to doubling the rate of vibration. Each note vibrates about 6% more rapidly than the note immediately below it. Every time one goes a certain distance along the keyboard, one multiplies the rate by a corresponding amount. This is just the same thing as happens on a slide-rule.

EXERCISES

1. If you can get hold of a slide-rule and a book of Logarithm Tables, verify the statement made in the text, that every number is marked on the slide-rule at a distance proportional to its logarithm.

2. Make a slide-rule for yourself, using tables of logarithms to tell you at what distance each number should be marked.

3. Make a slide-rule by the method explained in Chapter 6.

⌐ 4. Where is the square root of 10 marked on a slide-rule?

5. Check the accuracy of your slide-rule by finding on it 2 × 2, 2 × 3, 4 × 5 and other simple multiplications.

6. The logarithm of 2 is 0·301. The logarithm of 1·05 is 0·0212. How many years will money invested at 5% require to double itself?

7. An Eastern monarch sends 10,000 golden vessels to a brother monarch, whose kingdom is many days march distant. The gift is carried on camels. Each merchant, who supplies camels for some part of the journey, demands as commission 10% of what passes through his hands. Thus the first merchant hands over to the second not 10,000, but 9,000 golden vessels. Altogether, the vessels pass through the hands of twenty merchants. How many vessels does the brother monarch receive at the end?

CHAPTER 7

ALGEBRA – THE SHORTHAND OF MATHEMATICS

'Mathematics is a language.' – J. Willard Gibbs.

ALGEBRA plays a part in mathematics which may be compared to that of writing or of shorthand in ordinary life. It can be used either to make a statement or to give instructions, in a concise form.

Shorthand, of itself, does not make new discoveries possible. In the same way, most problems that can be solved by algebra can also be solved by common sense. Statements in algebra can be translated into ordinary speech, and vice versa. The statement in algebra is much shorter: some facts or instructions, which are easily written in algebraic form, are too long and complicated in ordinary speech. This is the advantage of algebra: while results *could* be got without it, it is unlikely that they *would*.

We shall consider some simple questions – perhaps not very

useful ones – to illustrate the form given to common-sense arguments, when the symbols of algebra are used.

The Cakes and Buns Problem

Most books on algebra, in a chapter headed 'Simultaneous Equations', deal with some such question as this. 'I visit a tea-shop on two occasions. The first time I order two buns and a cake; my bill is for 4d. The second time I order three buns and two cakes; the bill is for 7d. What are the prices of buns and cakes?'

I have tried this problem on people who know nothing about algebra, and they usually solve it. They argue: the second bill is 3d. more than the first. So 3d. represents the cost of the extra bun and cake. A bun and a cake cost 3d. But two buns and a cake cost 4d. So the difference, 1d., is the cost of a bun. A cake must cost 2d.

This problem may not sound important, yet in one form or another it repeatedly occurs in mathematical investigations of a very practical type. We shall ourselves need to solve such a problem in Chapter 8.

Mathematicians have therefore been forced to apply the argument outlined above on many different occasions. And – like other people – they have gradually introduced abbreviations to shorten the work. One can imagine the argument soon being written –

	2 buns & 1 cake	4d.
	3 buns & 2 cakes	7d.
So	1 bun & 1 cake	3d.
But	2 buns & 1 cake	4d.
So	1 bun	1d.
And	1 cake	2d.

Later one might begin to write 'b' where 'bun' comes, 'c' for 'cake'. If we replace '&' by '+', we have the modern form

$$2b + c = 4$$
$$3b + 2c = 7$$

So $\qquad b + c = 3$

But $\qquad 2b + c = 4$

So $\qquad b \quad\ = 1$

And $\qquad c = 2$

In this, b stands for the number of pence paid for a bun, c the number paid for a cake. You will notice that we write $2b$ for twice the number b. We do not write any multiplication sign between the 2 and the b. It is no use arguing whether a multiplication sign ought to be written here. If you feel happier with $2 \times b$, by all means write it that way. It is open to the objection that we often use x to represent a number, and $\times$ might easily be confused with x.

Of course $12b$ means *twelve* times b, not $1 \times 2 \times b$. You may feel it is confusing to have this distinction – but every shorthand system has its faults. In algebra, numbers such as 123 placed together have the same meaning as in arithmetic, but $2bc$ mean $2 \times b \times c$.

Try translating into ordinary language the following statements:

$$b + c + t = 6$$
$$2b + 3c + t = 11$$
$$4b + 8c + t = 23$$

b and c here have the same meanings as before, but each meal now contains a pot of tea costing t pence. This problem is also quite simple to solve. If you consider the difference in cost between the first meal and the second, you will obtain an equation which contains only b and c. Comparing the second meal with the third, you will get another statement, in which the price of tea does not appear. You now have two statements about buns and cakes:

$$b + 2c = 5$$
$$2b + 5c = 12.$$

If a bun and 2 cakes cost $5d.$, 2 buns and 4 cakes – twice as much – must cost $10d.$ So $2b + 4c = 10$. But we have above $2b + 5c = 12$. Comparing these, we see that $c = 2$. So $b = 1$. Going back to the first meal, we see that $t = 3$.

WORDS	PICTURE	ALGEBRA
I. Think of a number.		n
II. Add 6 to it.		$n+6$
III. Multiply by 2.		$2(n+6)$ or $2n+12.$
IV. Take away 8.		$2(n+6)-8$ or $2n+4$
V. Divide by 2.		$\dfrac{2(n+6)-8}{2}$ or $n+2$
VI. Take away the number you first thought of.		$\dfrac{2(n+6)-8}{2}-n$ or 2.
VII. The answer is 2.		$\dfrac{2(n+6)-8}{2}-n$ $=2.$

Each bag is supposed to contain as many marbles as
the number you thought of, whatever that was.

The same picture can often be described in different ways. Thus we
might describe the picture III as 'A bag and six marbles, twice,' or
simply as 'Two bags and twelve marbles.' In algebraic shorthand, these
descriptions are $2(n+6)$ and $2n+12$.

As a rule, there is no difficulty is solving problems of this type.

This shorthand can also be used to state truths. An old trick
runs as follows. 'Think of a number. Add 6 to it. Multiply by 2.
Take away 8. Divide by 2. Take away the number you first

thought of.' Whatever number you think of, the answer is always 2. Why?

We could deal with this by thinking in pictures. You think of a number – any number. We will think of this as marbles placed in a bag. 'Add 6 to it.' This gives us one bag and six loose marbles. 'Multiply by two' – two bags and twelve marbles. 'Take away 8' – two bags and four marbles. 'Divide by 2' – one bag and two marbles. 'Take away the number you first thought of' – that is, take away the bag. Two marbles remain – whatever the number in the bag.

In algebra, we need not talk about bags of marbles. We say, let n stand for the number you think of. 'Add 6'; we get $n + 6$. 'Multiply by 2', $2n + 12$. 'Subtract 8', $2n + 4$. 'Divide by 2', $n + 2$. 'Take away n', 2 is the answer.

We can express the whole process, and the fact that the answer is always 2, by writing the single equation

$$\frac{2(n + 6) - 8}{2} - n = 2.$$

The expression on the left-hand side indicates that you take twice $(n + 6)$, subtract 8, and divide by 2, finally subtract n. One line of symbols replaces a paragraph of talk.

This example shows that two apparently different expressions may in fact represent the same thing. An important part of algebra therefore consists in learning how to express any result in the simplest possible way: this is known as Simplifying.

It is sometimes possible for a question to have two answers which at first sight appear different, but which are actually both correct.

Suppose, for instance, you were asked to discover the rule by which the following numbers have been chosen: 0, 3, 8, 15, 24, 35, 48, 63. You might notice that these numbers are given by the rule 1^2-1, 2^2-1, 3^2-1, 4^2-1, 5^2-1, 6^2-1, 7^2-1, 8^2-1. (You will remember from Chapter 6 that 5^2 is short for 5×5.) In short, the nth number is n^2-1.

But you might also notice that 63 is 7×9, that 48 is 6×8, and so on. The eighth number is the number before 8 (that is, 7) multiplied by the number after 8 (that is, 9). The first number, 0, is the number before 1 (i.e., 0) multiplied by the number after 1 (i.e., 2). This suggests the rule for the nth number: multiply the number before n (which is $n - 1$) by the number after n (which is $n + 1$). This gives us the formula $(n - 1)(n + 1)$.

Both these rules are correct. Whatever number you may choose for n, you will always find that $(n - 1)(n + 1)$ is the same as $n^2 - 1$.

Here we have used algebraic signs as a shorthand for writing *instructions*, how to find the numbers in a certain set. This use of algebra is very common. The person who uses a formula need not understand why a formula is right. For instance, a sapper who has to blow up a railway bridge will work out how much explosive to use by means of a formula: he does not need to know how the formula is obtained in the first place. In the same way, there is a rule which says, if you wish to see n miles out to sea, you must have your eyes $\dfrac{2n^2}{3}$ feet above sea-level. This formula is found by means of geometry: but without knowing any geometry you can use this formula, and discover that a tall man on the beach can see nearly 3 miles out to sea (since $\dfrac{2 \times 3^2}{3}$ gives 6 feet as the height needed), while to see 12 miles, you need a cliff 96 feet high. Such a formula could be used in designing a battleship, to tell how high an observer must be to see the effect of the ship's guns.

Most formulae contain several different letters. For instance, we might wish to know how much metal is required to make a circular tube. We must be told how long the tube is, how thick the wall is, and the measurement around the tube. Let us call the *length* L inches, the *thickness* T inches, the outside *measurement* around the tube M inches. A formula tells us the tube will then contain $LT(M - 3 \cdot 14T)$ cubic inches of metal. Thus a tube 10 inches long, $\frac{1}{2}$ inch thick, and 15 inches round will contain $10 \times \frac{1}{2} \times (15 - 3 \cdot 14 \times \frac{1}{2}) = 5 \times 13 \cdot 43 = 67 \cdot 15$ cubic inches. A rule such as this could be stated in words, but would be much longer. The shorthand is so simple - L for length, T for thickness,

M for measurement – that no difficulty can arise in learning it. Yet many people are terrified by the sight of a page of algebraic symbols, and others have a reputation for immense intelligence because they understand algebra.

In bygone ages, a man who could read and write counted as a scholar. Today we think nothing of reading and writing. Algebra too is a language – neither more nor less mysterious than ordinary print, once its alphabet and its grammar have been learned.

EXAMPLES

1. We may express the instructions, 'Think of a number (n), double it, and add 5' by the shorthand sign $2n + 5$. Translate into algebraic shorthand the following sentences –

   (i) Think of a number, add 5 to it, and double the result.
   (ii) Think of a number, multiply by 3, and add 2.
   (iii) Think of a number. Write down the number after it. Add the two numbers together.
   (iv) Multiply a number by the number after it.
   (v) Think of a number, and multiply it by itself.

2. Translate the following shorthand signs back into sentences such as those of Question 1.

   (i) $4n + 4$.   (iii) $3(n + 1)$.   (v) $\frac{1}{2}n(n - 1)$.
   (ii) $n - 1$.   (iv) $\frac{1}{2}(4n + 8)$.

Work out what these signs would give if the number thought of, n, happened to be 6. Also what they would give if the number n was 3.

3. To see n miles out to sea you must have your eye $\dfrac{2n^2}{3}$ feet above sea-level. Make a table showing how high your eye would have to be, in order to see 0, 1, 2, 3, ... 10 miles.

4. We have seen that $(n - 1)(n + 1)$ gives exactly the same number as $n^2 - 1$. We tested this by putting n in turn equal to 1, 2, 3, ... 8. This did not prove that the two expressions were

always the same, but it made it likely. By this test you can show that some of the statements below are *probably true*, while others are *certainly untrue*. Which are which? Here n means 'any number.'

(i) $(n + 1) + (n - 1) = 2n.$

(ii) $n = 2n.$

(iii) $2(n + 3) = 2n + 6.$

(iv) $n^3 - 1 = (n - 1)(n^2 + n + 1).$

(v) $4n + 2$ is always an even number (n being a whole number).

(vi) $n(n + 1)$ is always an even number (n being a whole number).

(vii) $(n + 1)(n + 1) = n^2 + 1.$

(viii) $\dfrac{2n}{n^2 - 1} = \dfrac{1}{n + 1} + \dfrac{1}{n - 1}.$

5. If it takes a minutes to set up a certain machine, and b minutes to make a single article on the machine, how long does it take to set up the machine and make 10 articles?

How long would it take to set up the machine and make n articles?

Write down in turn the time needed to set up the machine; the time needed to set it up and make 1 article; the time to set it up and make 2 articles, etc.

6. In question 5, the time taken to make n articles after setting up the machine is $a + bn$ minutes. The last part of the question requires us to put n in turn equal to 0, 1, 2 ... in the formula $a + bn$, giving a, $a + b$, $a + 2b$, ... etc. Putting a definite value for n in the formula is called *substituting* for n.

Thus, $a + 2b$ is the result of substituting $n = 2$ in the formula $a + bn$.

Again, if in the formula $n^2 - 1$ we substitute $n = 7$, we get $7^2 - 1$. In question 3, we substituted in turn 0, 1, 2 ... for n in the formula $\dfrac{2n^2}{3}$. You will be able to follow the argument of Chapter 8 only if you are familiar with this idea. That is the point of the following examples.

(i) If a train travels at v miles an hour, it goes nv miles in n hours.

Make a table showing how far it goes in 0, 1, 2, 3, 4 and 5 hours.

(ii) What does this table become if $v = 30$, i.e. if the train travels at 30 m.p.h.?

(iii) What does it become if $v = 10$?

(iv) What is the result of substituting $n = 4$ in the expression $5n - 1$?

(v) What is the result of substituting $n = 1$ in $2n^2 + 3n + 5$?

(vi) If you substitute in turn, 0, 1, 2, 3, ... etc., for n in the formula $n^2 + 4n + 4$, what do you notice about the answers?

(vii) What is the result of substituting in turn $n = 0, 1, 2, ...$ in the expression $an^2 + bn + c$?

Note – This last question seems to trouble students, but we need the answer to it for Chapter 8. Compare it with section (v). Putting $n = 1$ in $2n^2 + 3n + 5$ gives 10. For when $n = 1$, n^2 becomes 1^2, that is, 1, and the expression $2n^2 + 3n + 5$ becomes $2 \times 1 + 3 \times 1 + 5$, that is, $2 + 3 + 5$. So that when $n = 1$, the three numbers that occur in the expression (2, 3, and 5) *are simply added together*. This would happen whatever these numbers were. If you put $n = 1$ in the expression (say) $10n^2 + 17n + 35$, you would get the result $10 + 17 + 35$, which adds up to 62. If you put $n = 1$ in any expression of this type (so much n^2, so much n, and a number added together), you get the answer by adding together the three numbers that occur in the expression. So, if you put $n = 1$ in $an^2 + bn + c$ you get the answer $a + b + c$.

In the same way, you will find that the result of putting $n = 0$ is simply to give you the third number.

For instance, when $n = 0$ the expression $4n^2 + 173n + 45$ becomes 45.

When we substitute $n = 0$, $an^2 + bn + c$ becomes c.

You will also find that the result of substituting $n = 2$ is to give you

4 times the first number in the formula.
+2 times the second number in the formula.
+   the third number in the formula.

In shorthand, this result is $4a + 2b + c$.
Find for yourself the rules for what you get when you put
$n = 3$, or $n = 4$, or $n = 5$.

If you find great difficulty with this chapter, try to get hold of an engineer, who will tell you just what he does when he uses a formula for a practical problem. The earlier part of Chapter 8 may also help you to see how a formula is used in practical life. It is only towards the end of Chapter 8 – 'The Calculus of Finite Differences' – that we have to do problems similar to Question 6, above.

CHAPTER 8

WAYS OF GROWING

'*When it is considered how essential is their use in a vast range of trades and professions – from plumbing to* Dreadnought *building – it is hardly extravagant to say that facility in the working, interpretation and application of formulae is one of the most important objects at which early mathematical studies can aim.*'
– T. Percy Nunn, *The Teaching of Algebra*.

WE are very often interested in knowing how quickly a thing grows. If one skyscraper is twice as high as another, it must have a stronger framework to support the extra weight. It will cost more than twice as much to build. How much more? Four times? Eight times?

As an army advances into hostile territory, the difficulty of bringing up supplies and maintaining communications increases. To cover 1,000 miles requires more than ten times the lorries necessary for 100 miles: how many, then?

If you are fire-watching on a high building, you may need to jump off in an emergency. Is it twice as dangerous to jump from 40 feet as from 20? Or more than twice? Or less than twice?

If a housewife is buying firewood, and pays 6d. for a bundle measuring 1 foot around, how much should she pay for one measuring 6 inches?

In all these questions, we are interested in the different ways in which something grows. The answer may be quite important for practical purposes. Anyone who tries to provide housing economically has to know the answer to the first question: otherwise he may find that a saving in ground-rent has been altogether swallowed up in extra costs for building material.

Again, many would-be inventors have been disappointed through ignorance of the effects of change in scale. Many people have, in the past, invented flying-machines and successfully made small models which flew very well. They then enlarged their model, and built a full-size machine – and it would not fly at all. The reason was that the weight of a machine and the lifting-power vary in entirely different ways. If one constructed a flea, enlarged to the size of an elephant, its performance would be entirely different from that of an actual flea – as you can well imagine.

It is therefore quite natural that engineers and scientists expect mathematicians to supply them with a simple way of writing down the manner in which any quantity grows.

Of course, things grow in a great variety of ways. If, for instance, one considers the population of Manchester during the last 150 years, this is a quantity which has changed in a very complicated way. A mathematician can study it, and describe it, but you must not expect the description to be very short or simple. Many things will enter into it – the Industrial Revolution, the varying fortunes of the cotton trade, evacuation of children in war-time, and so on. On the other hand, some things vary in a very simple way. We shall give examples below. In between these two extremes come a great variety of types of growth, which can be studied and written down with greater or less difficulty. You must not expect mathematics to make a complicated question simple. Mathematics may help us to discover the underlying causes of things, but if the causes are many and intricate, then the mathematical description too will be far from simple. We shall here study simple cases only: do not make the mistake of supposing that every problem, however profound, can be forced into these simple forms.

The Simplest Form of Growth

An example of a simple relation is the cost of any article bought by the yard. If one yard of lath costs 2*d*., 2 yards will cost 4*d*., *x* yards will cost 2*x* pence. If *p* stands for the price in pence of *x* yards, we have $p = 2x$.

We can make this formula more general. A yard of lath need not

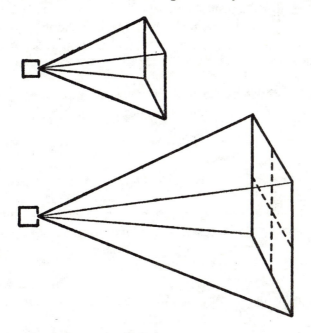

THE SIZE OF CINEMA SCREENS

In the lower diagram the screen is twice as far from the projector as in the upper diagram. The lower screen is twice as broad and twice as high as the upper one. The dotted lines show that the lower screen could be cut into four screens, each as large as the upper screen.

cost $2d$. Suppose it costs a pence, where a may stand for any number. Then the price of x yards is given by $p = ax$. We assume that the price of each yard does not depend on the number of yards bought: there is to be no reduction for quantity. In mathematical language this is stated by saying that a is *constant*.

Such relations are common. For instance, the circumference and diameter of a circle – C and D – are connected by the relation $C = 3 \cdot 14\ D$. Again, in a spring-balance the spring stretches by a distance proportional to the weight hung on it. If 1 lb. causes the spring to stretch k inches, than x lb. will cause it to stretch kx inches. This fact was discovered by Hooke, about 1660. Hooke was led to study the properties of springs by his work on clock-making. His invention of the balance-wheel – by which a hair-spring replaces the pendulum of a clock – was a practical consequence of his investigation. Hooke's Law is true only for reasonably small weights. A very heavy weight will stretch a spring too far: when the weight is removed the spring does not return to its original length.

Other formulae of the type ax will be found in almost every branch of mathematics, engineering or science.

Powers of x

Another type of growth occurs when a cinema projector, or magic lantern, is moved further away from the screen. If you double the distance of a magic lantern from the screen, the picture will take up – not twice, but – four times as much space. If you treble the distance, the amount of material needed for the screen will be nine times as much.

The rule is clear: 4 is 2×2, or 2^2, 9 is 3×3, or 3^2. If we put the lantern n times as far away, we need n^2 times as much material for the screen.

In the same way, if we enlarge a photograph or a map n times, n^2 times as much paper will be necessary.

We have here the answer to our earlier question about firewood. The bundle tied with a 12-inch string contains four times as much firewood as the 6-inch bundle. You can see that the large

Fig. 5

bundle must contain more than twice the amount of the small one, by looking at Fig. 5. The black part represents two small bundles: the large circle represents one large bundle.

If a stone is dropped from the top of a cliff, you will find that it falls about 16 feet in the first second; after two seconds it has fallen 64 feet; after three seconds, 144 feet. These results can be expressed by the formula that, in x seconds, the stone falls about $16x^2$ feet. Thus in 5 seconds the stone would fall 16 times $5^2 =$ 16 times $25 = 400$ feet. If you can remember this formula, and have a watch with you, it is easy to find the height of a cliff, or the depth of a well, by dropping a stone, and noting how many seconds it takes to reach the bottom.

Another formula gives the speed with which it strikes the bottom. This is $v^2 = 64\,h$. In this formula h is the height of the cliff in feet, and v is the velocity, in feet per second, with which the stone lands. If the cliff is 100 feet high, h is 100, so $v^2 = 6,400$, and $v = 80$. To produce a speed twice as big, the cliff would have to be *four* times as high. (This answers the question about fire-watching.) It is easily worked out that a speed of 3 feet a second is roughly the same as 2 miles an hour. So 80 feet a second is about 53 miles an hour. Jumping off a house 100 feet high is as dangerous as a car crash at 53 miles an hour.

In the same way, we could discuss the way of growing represented by x^3. You would need 8 cubes of sugar to make 1 cube with twice the normal measurements. To make a cube x times the usual size, you would need x^3 ordinary cubes. If you enlarge any solid object x times – it need not be a cube – you multiply the material contained by x^3. For instance, if you double all the measurements of a drawer, or a box, or a trunk, you multiply by 8 the amount that can be contained.

When a model of any object is enlarged, all these different kinds of growing may be involved. Suppose, for simplicity, we have a box in the form of a cube, made out of cardboard. If the side of the cube is 1 foot, the box will hold 1 cubic foot, 6 square feet of cardboard will be sufficient to make it, and a string 4 feet long could extend round it. If we make instead a box in the form of a cube, but with each side 2 feet long, we find that it holds eight times as much, but that it needs only four times as much cardboard, and a piece of string only twice as long can extend round it. It is cheaper to pack goods in large boxes than in small ones – so long as the cardboard does not burst.

It is easy to see why the early aeroplane inventors had disappointing results, when they tried to enlarge the scale of their models. If the scale is doubled, the weight is multiplied by 8, but the wing-surface is multiplied by only 4.

x, x^2, and x^3 thus turn up naturally enough, when we consider changes in the scale of models and plans. In other applications, we make use of x^4, x^5, x^6 and so on.

For instance, a common device for storing energy is a flywheel. Suppose we make two flywheels by cutting circular pieces out of a sheet of metal – one circle having twice the radius of the other. Suppose both wheels are turning at the same rate – say, one revolution a second. Will the large wheel have twice, or four times, or eight times the energy of the small one? No. Experiment shows that it has *sixteen* times as much, 2^4 as much, that is. If we enlarge the radius x times, we increase the energy x^4 times. Aeroplane engines can be started by means of a flywheel. The flywheel is made to rotate by hand, and is then suddenly connected to the aeroplane engine. If you were using the larger flywheel, mentioned above, you would have sixteen times the energy at your disposal, as compared with a man using the small one: as it would take you sixteen times as long to get your flywheel spinning, you would obtain a vivid picture of the meaning of 2^4!

If you double not only the radius of the flywheel, but also the thickness of the metal, you multiply the energy by 2^5, or 32. In this case, the bigger flywheel is twice as large *in every direction* as the smaller one. The effect of enlarging a flywheel x times in every direction, is to increase its energy (at a given speed or rotation) x^5 times.

x, x^2, x^3, x^4, x^5 are called the first, second, third, fourth and fifth powers of x. Instead of 2, 3, 4 or 5, we could have any number. Using n as an abbreviation for 'any number', we may say that x^n is called the nth power of x.

Powers of x may occur mixed with each other, and with constants. For instance, a tennis club might charge 5*s*. entrance fee, and 1*s*. for every afternoon on which a member actually played during a season. The cost of one afternoon's play would thus be 6*s*., of two would be 7*s*., of x would be $(5 + x)$ shillings.

Again, if a ball is thrown straight up with a speed of 40 feet a second, after x quarter seconds its height is given by the formula $10x - x^2$ feet. (Quarter-seconds are used instead of seconds partly to give simpler numbers, partly because a ball spends such a short time in the air.) We could make a table as follows –

Number of quarter-seconds.	0	1	2	3	4	5	6	7	8	9	10
Height in feet.	0	9	16	21	24	25	24	21	16	9	0

Examine the set of numbers in the lower row of this table. You will notice that it reads the same backwards as forwards. Do you notice anything else about it?

Write down the change between each number and the next one. Thus –

Numbers.	0	9	16	21	24	25	24	21	16	9	0
Change.		9	7	5	3	1	−1	−3	−5	−7	−9

There is a very simple rule to be seen in this set of numbers: each number is 2 less than the number before it. This is due to the steady downward drag which gravity exerts on the ball. In the first interval of a quarter-second the ball rises 9 feet. But it is being slowed down all the time. In the second interval it covers only 7 feet, in the third 5 feet, and so on. In the fifth it rises only 1 foot. In the sixth it descends 1 foot. (As usual, $+$ 1 means one foot *up*, $-$ 1 one foot *down*.) Now it comes down faster and faster: 3, 5, 7, 9 feet in successive intervals.

We can keep on writing down such rows of numbers – each row giving the changes in the row above. We should thus get the following table –

TABLE I

0		9		16		21		24		25		24		21		16		9		0
	9		7		5		3		1		−1		−3		−5		−7		−9	
		−2		−2		−2		−2		−2		−2		−2		−2		−2		
			0		0		0		0		0		0		0		0			
				0		0		0		0		0		0		0				

All the numbers in the third row are the same, −2. There is no change as we go from one to the next. So all the numbers in the fourth row are noughts. We can go as far as we like: we shall merely get more noughts, in the fifth, sixth and following rows.

Let us try this on some other expressions we have already had. If we write down the numbers corresponding to x^2, we obtain –

TABLE II

0		1		4		9		16		25		36		49		64		81
	1		3		5		7		9		11		13		15		17	
		2		2		2		2		2		2		2		2		
			0		0		0		0		0		0		0			

If we try x^3 we find

TABLE III

0		1		8		27		64		125		216		343		512
	1		7		19		37		61		91		127		169	
		6		12		18		24		30		36		42		
			6		6		6		6		6		6			
				0		0		0		0		0				

Try for yourself x^4 and x^5. Make up expressions such as $2x + 3$, $x^2 + 5x + 7$, and try it on these. You will find that, after a certain number of rows, you always get noughts. What is the rule giving the number of rows that occur, before the noughts are reached? The answer to this question is given later. But try to guess it for yourself. Work out a large number of different examples: group together those which have one row, then noughts: in another group put those with two rows, and so on. The rule is quite a simple one.

Exponential Functions

We have just seen that any expression, made up by mixing together powers of x, will lead to rows of noughts at a certain stage of the process described above.

Not all ways of growing have this property: in fact, expressions formed by mixing powers of x are the only type that possess this characteristic.

If you try some other rule, you will soon see that this is true. Take, for instance, the set of numbers 1, 2, 4, 8, 16, 32, 64, 128, etc., where each number is formed by doubling the previous one. This set corresponds to the formula 2^x. (Remember that we start with $x = 0$. If a sum of money doubled itself each year, a high rate of interest, £1 would become £2 after 1 year, £4 after 2 years, etc. After x years it would be £2^x.) If we make a table for this set of numbers, using the same method as before, we get –

```
1   2   4   8   16   32   64   128  . . . . .
    1   2   4   8   16   32   64   . . . . .
        1   2   4   8   16   32   . . . . . .
```

Each row is exactly the same as the one before! However long we go on, we shall never find a row all noughts.

Try 3^x. This gives us the table –

```
    1   3   9   27   81   243   . . . . .
        2   6   18   54   162   . . . . .
            4   12   36   108   . . . . .
```

Here each row is *twice* the previous one. However long we go on, we shall never find a row all noughts.

2^x, 3^x are called *exponential functions*. If we use a as shorthand for 'any number', a^x is an exponential function.

The Calculus of Finite Differences

It very often happens that we want to know the rule by which a certain set of numbers has been formed. An engineer might find, by experiment, the pressure required to burst boilers made from sheet metal of various thicknesses. It would be helpful to other engineers if he could express his results in the form of a simple rule. A scientist might measure the size of a plant each day, and try to find the rule by which it grew.

A great part of science consists of the attempt to find rules by studying the results of experiments.

When one quantity depends on another, it is said to be a *function* of the latter quantity. Thus, the bursting pressure of a boiler depends on the thickness of the boiler walls. Calling the pressure p and the thickness t, we say that p is a function of t; the pressure needed to burst a wall of thickness t may be written $p(t)$. Thus $p(2)$ would mean the pressure needed to burst a boiler built with metal 2 inches thick, $p(\frac{3}{4})$ would mean the pressure to burst a boiler with $\frac{3}{4}$-inch walls. Naturally, we suppose that the design of the boiler is fixed, and that the same metal is used for all the experiments.

In the same way, if we denote 'the number of days' by x, and 'the size of a plant in inches' by y, y is a function of x. $y(17)$ will mean the size of the plant after 17 days: $y(x)$ the size after x days.

If we say, 'What function is y of x? we mean, 'By what particular rule is y connected with x?'

This question is used in intelligence tests. A child is shown the numbers 1, 2, 3, 4, 5 and is asked, 'What is the next number?' Of course 6 is the answer. An older child might be shown 2, 4, 6, 8 and expected to guess the next number as being 10.

Such simple cases can be guessed without any special method. But suppose you were shown the table below –

x	0	1	2	3	4	5	6	7	8	9	10
y	1	3	7	13	21	31	43	57	73	91	111

How is the number y in the second row found, corresponding to any number x in the first row? A child might be forgiven if, shown the numbers 1, 3, 7, 13, 21, 31 it failed to guess that the next number was 43! But how quickly our method – of writing down the change from one number to the next – supplies a clue. It gives the table –

TABLE IV

1		3		7		13		21		31		43		57		73		91		111
	2		4		6		8		10		12		14		16		18		20	
		2		2		2		2		2		2		2		2		2		2	
			0		0		0		0		0		0		0		0		0		

We have noughts in the bottom row: the formula must be a simple one, containing only powers of x.

But how many powers of x shall we need? Shall we have to bring in x^5 as was necessary in the case of the flywheel? Or need we not go so far?

Perhaps you have already discovered the answer to the question asked earlier.* If not, here is the answer, Any formula, such as $2x + 3$, which contains no power higher than x, gives us two rows of numbers, and then noughts. If x^2 comes in – as in $5x^2 + 3x - 2$, for instance – we have three rows, then noughts. If x^3 comes into the formula, we have four rows, then noughts. And so on. If x^n occurs, we have $(n + 1)$ rows before the noughts come. This works also the other way round. If we have four rows before the noughts, it will be possible to find a formula which does not use any power above x^3. If we have $(n + 1)$ rows, the formula will contain powers up to x^n.

This helps us to find the formula which gives the numbers 1, 3, 7, 13, etc. This set of numbers, as we have just seen, leads to a table containing only three rows. The formula cannot contain any power of x higher than x^2. It will be sufficient to take a certain amount of x^2, together with a certain amount of x, and with some number added. In algebraic shorthand, our formula will be $ax^2 + bx + c$, where a stands for the number that goes with x^2, b for the number that goes with x, c for the number that is added. (Thus, in the formula $5x^2 + 3x - 2$, a is 5, b is 3, c is —2.) We do not yet know what a, b and c have to be. All we

* Page 94

know is that it is possible to get the right formula by choosing the proper values for a, b and c. This of course is a great help. When we started this problem we had to be prepared for *any* formula: it might have been $x + 2^x$, or x^9, or even worse expressions.

Once we know that the formula is of the type $ax^2 + bx + c$, it is very easy to find a, b and c. We know that the numbers 1, 3, 7, 13, etc., result if, in the proper formula, we replace x by the numbers 0, 1, 2, 3, etc., in turn.

If in the formula $ax^2 + bx + c$ we replace x by 0, we get c: if we replace it by 1, we get $a + b + c$: if we replace it by 2, we get $4a + 2b + c$. (If you like, you can turn these results into words by reading $4a$ as 'four times the number that goes with x^2 in the formula', and so on.)

We can now compare these two sets of results. If y is given by the formula $ax^2 + bx + c$, $y(0) = c$. But $y(0)$, the value of y corresponding to $x = 0$, is 1. c must be 1. Again the formula gives $y(1) = a + b + c$. But $y(1)$ is 3. So we must choose a, b and c in such a way that $a + b + c = 3$. In the same way, comparing the formula for $y(2)$ and what it ought to give, we get the equation $4a + 2b + c = 7$. Altogether we have three equations:

$$c = 1$$
$$a + b + c = 3$$
$$4a + 2b + c = 7.$$

This is exactly similar to the problem of the cakes and the buns and the pot of tea. It is easily solved by the method described in Chapter 7, and leads to the result $a = 1$, $b = 1$, $c = 1$. So that we have the formula $y = x^2 + x + 1$. This is the rule by which the numbers in the table were found.

In the subject known by the imposing name of the Calculus of Finite Differences, the method we have just used is further developed, and proofs of its correctness are given.

It has been found convenient to introduce certain abbreviations. We have had to keep on referring to 'the second row in the table', 'the third row', and so on. To avoid this, certain signs are used, as names for these rows. The first row (which, in our last example, contained the numbers 1, 3, 7, 13 ...) we have already called y. The second row (the numbers 2, 4, 6, 8...in that example) are called Δy. The sigh Δ is short for 'the change in'. As each row represents the changes that occur in the previous row, we get one more sign Δ every time we go down a row. The third row, for instance, represents the changes in Δy, and could be written $\Delta\Delta y$. Note that Δ does not stand for any number as did a, b and the other letters. Δ stands for '*the change in*' – that and nothing else. It can always be replaced by these words. Usually $\Delta\Delta y$ is still further shortened to $\Delta^2 y$. This is especially convenient when large numbers of Δ are involved. For instance $\Delta^5 y$ is much more convenient than $\Delta\Delta\Delta\Delta\Delta y$ as an abbreviation for the numbers in the sixth row.

Sometimes we want to refer briefly to a particular number in one of the rows. We have already used the sign $y(x)$ to describe the number in the first row which corresponds to the value x: so that the numbers of the first row are $y(0)$, $y(1)$, $y(2)$, $y(3)$ and so on. We use similar signs for the numbers in the following rows. The numbers in the second row are called $\Delta y(0)$, $\Delta y(1)$, $\Delta y(2)$ and so on; in the third row, $\Delta^2 y(0)$, $\Delta^2 y(1)$, $\Delta^2 y(2)$ etc.; and so on for any row.

You will find these signs in any book on the Calculus of Finite Differences. At first they may seem strange, but once you are accustomed to them, and have realized that $\Delta^2 y(1)$ means nothing more terrifying than 'the second number in the third row' of a table such as Table III or Table IV, you will find the subject quite a good one to experiment with. You may try your hand at the following problems.

(1) A car drives past a lamp-post. One second later it is 3 yards away from the lamp-post; after 2 seconds, 10 yards; after 3 seconds, 21 yards; after 4 seconds, 36 yards. How far away is it after $\frac{1}{2}$ second, $1\frac{1}{2}$ seconds, $2\frac{1}{2}$ seconds? Is it speeding up or slowing down?

(2) What is the missing number in the following set?

$$3, 4, \ldots 24, 43, 68.$$

If you succeed in guessing the right number, the table for Δy, $\Delta^2 y$, etc., will make it quite clear that you are right. You will not be in any doubt about it, once you have tried the right number. And it must be a number between 5 and 23. If the worst comes to the worst, you can try all of these in turn.

Binomial Coefficients

It is possible to work out a table similar to Table IV, if we suppose the first row, y, to contain *any* set of numbers. In fact, we may represent the numbers in the first row by algebraic symbols. Let a stand for the first number (whatever that is), b for the second, c for the third, d for the fourth and so on. The row y then reads –

$$a, b, c, d, e, f \ldots$$

How are we to form the next row Δy? The first number in it shows the change from a to b. It is obtained by subtracting a from b, and may therefore be written $b - a$. In the same way, the next number may be written $c - b$. (Check these statements for yourself. In Table IV, what numbers are a, b, and c? Is it true that the row Δy begins with numbers equal to $b - a$ and $c - b$?) The second row in fact is $b - a$, $c - b$, $d - c$, $e - d$, $f - e$, etc.

From the second row the third row can be found. The first number in it is $(c - b) - (b - a)$, which simple algebra shows to be equal to $c - 2b + a$. The second number in this row is $d - 2c + b$.

Continuing in this way, we obtain the expressions collected in Table V.

TABLE V

a		b		c		d		e
	$b—a$		$c—b$		$d—c$		$e—d$	
		$c—2b+a$		$d—2c+b$		$e—2d+c$		
			$d—3c+3b—a$		$e—3d+3c—b$			
				$e—4d+6c—4b+a$				

You will notice certain things about this table. A particular set of numbers appears in each row. In the row $\Delta^4 y$, for instance, we have the numbers 1, 4, 6, 4, 1. In the row $\Delta^3 y$ we find the numbers 1, 3, 3, 1. In the row $\Delta^2 y$ we find 1, 2, 1, and in the row Δy we find simply 1, 1. (No notice has been taken of whether these numbers appear with a + or — sign.) You will notice that these sets of numbers read the same backwards as forwards. For example, 1, 3, 3, 1 is the same, whether read backwards of forwards. You will notice that the first and last numbers are always 1. What else do you notice? What is the rule that gives the number next to the end? Can you find the formula for the number next to that? (You will need to work out a few more rows of Table V to do this.) Apply the method already explained, for finding the formula of a set of numbers.

These numbers are known as the *Binomial Coefficients*. Mathematicians came to know them in exactly the way you have done – by noticing that they turned up in the course of work. They turn up, for instance, if you work out 11^2, 11^3, and 11^4, which are, in fact, 121, 1331, 14641. (After this point *carrying* comes into the arithmetic, and the simple connexion does not hold. The numbers in $\Delta^5 y$ are 1, 5, 10, 10, 5, 1, and of course the number *ten* cannot appear as a single figure in 11^5. 11^5 is actually 161,051, which does not read the same backwards as forwards.) They appear, too, in $(x + 1)^2$, $(x + 1)^3$ and $(x + 1)^4$, etc. We may write them in a table, as below –

TABLE VI

```
1 1
1 2 1
1 3  3  1
1 4  6  4  1
1 5 10 10  5  1
1 6 15 20 15  6 1
1 7 21 35 35 21 7 1
```

Now you can measure your place among the great mathematicians. This table was known as early as 1544. Gradually people noticed all sorts of odd facts about it. But it was not until 1664, 120 years later, that the greatest of English mathematicians found *a formula* giving the numbers in any column of Table VI. The first column is obvious enough – always 1. The second column contains 1, 2, 3, 4, 5, 6, 7 – a simple law here. But what is the rule for the third column, 1, 3, 6, 10, 15, 21? You will find this problem quite easy if you use the method outlined earlier in this chapter – make a table on the lines of Tables I-IV, and then hunt for a formula.

The rule that Newton found (and that I hope you will find for yourself) is known as the *Binomial Theorem*. That is all the Binomial Theorem is – a rule for writing down the numbers in Table VI.

The object of explaining Δy, $\Delta^2 y$, etc., to you is that you may see how theorems are discovered, and may be able to discover results for yourself.

EXERCISES

1. If the temperature of L feet of steel is raised T degrees Fahrenheit, the length of the steel increases by an amount $0\cdot000006\ LT$.

If the temperature of a mile of steel (say, on a railway) rises 10 degrees Fahrenheit, how much extra room will it need?

2. In scientific work temperature is measured in degrees Centigrade. This can be changed to degrees Fahrenheit by means of the formula

$$F = \frac{9C}{5} + 32,$$

where F degrees and C degrees stand for the temperature Fahrenheit and Centigrade respectively. What temperature Fahrenheit corresponds to 15 degrees Centigrade? What is the temperature 90 degrees Fahrenheit in terms of degrees Centigrade?

3. 10 feet of common 2-inch-bore lead piping weighs 50 lb. What is the formula for the weight of x feet of such lead piping?

4 One square foot of ordinary brickwork will support with

safety a weight of 6 tons. How many tons will a piece of brick-work x feet square support?

With particularly good bricks, 1 square foot of brickwork will support 54 tons. What weight will a block x feet square support?

A brickwork foundation capable of bearing a weight of 1,000 tons is required. The foundation is to be square in shape. How large must it be if it is made (i) of ordinary brick, (ii) of particularly good bricks?

5. When a train moves at x miles an hour, the total pressure on the front of the locomotive, due to air pressure (y lb.), is given by the following table.

x	0	20	40	60	80	100
y	0	79·5	318·0	715·5	1272·0	1987·5

Is there a simple formula connecting x and y? If so, what is it?

6. One sometimes sees on trams a table giving the fare between any two points on the route, as in the illustration.

From \ To	Blue Boar	Green Man	Black Market
Green Man	1*d*.		
Black Market	2*d*.	1*d*.	
Red Lion	3*d*.	2*d*.	1*d*.

It is clear that no diagram at all is necessary unless the tram has at least two stations to connect. With 3 stations, the diagram contains 3 squares. Work out the number of squares in the diagram when there are 4, 5, 6, etc., stopping stations. How many squares are needed if there are x stations? Is there any connexion with the numbers in Table VI?

7. If n teams enter for a knock-out competition, how many matches will be played? (See the question at the end of Chapter 5.)

8. In Britain and the United States the horse-power rating of a motor-car is found from the formula –

$$H = \frac{2Nd^2}{5}$$

where

$H =$ horse-power rating

$N =$ number of cylinders

$d =$ bore of cylinder in inches.

What bore would be necessary, for a four-cylinder car, to give 40 H.P.? What for 10 H.P.? What would give a horse-power just below 23 H.P.?

9. The breaking strength of three-strand Manila rope is given by the formula –

$$L = 5000 \ d(d+1)$$

where L lb. is the load needed to snap a rope d inches in diameter. How many lb. will be required to snap a rope $1\frac{1}{2}$ inches in diameter? What rope will just snap under a load of 60,000 lb.?

10. The safe load, S hundredweight, for a rope measuring C inches in circumference is given by the formula –

$$S = C^2.$$

How many hundredweight may safely be put on ropes of circumference 2, 3, and 4 inches? What size of rope is needed to carry safely $\frac{1}{4}$ ton? How many ropes 2 inches in circumference would be needed to support $\frac{1}{4}$ ton? (Fractions of ropes are not used.) How many ropes 3 inches around would do the same job?

Further formulae, on which to practise, can be found in Kempe's *Engineer's Yearbook*, from which many of the examples used here have been drawn. Formulae will be found covering subjects varying from the amount of sludge deposited by sewage to the amount of rainfall in Northern India.

CHAPTER 9

GRAPHS, OR THINKING IN PICTURES

*'Care is taken, not that the hearer may understand, if he will,
but that he must understand, whether he will or not.'*
— Henry Bett, *Some Secrets of Style.*

THE great problem of every teacher is to present facts in such a
way that the students cannot help *seeing* what is meant. A bald
statement is soon forgotten. Vivid images remain in the memory.
Many people must have noticed the difference between reading
a history text-book and seeing an historical film. Whatever the
relative accuracy of the book and the film, the film certainly makes
one realize events more intensely, and remember them longer.

In films it is sometimes necessary to explain quite complicated
ideas, not to a class of students, but to an audience which represents the whole population of a country. Cinema audiences, too,
are in no mood for concentrated thought. They want to relax, to
be amused. It is extremely instructive to examine how film
directors go about the job. They rarely fail to make their point
understood – a fact which should be seriously considered by those
defeatists in education whose perpetual alibi is the stupidity of
pupils.

In our last chapter we considered the different ways in which
any quantity can grow. Suppose we wished to present this idea
to a cinema audience. How could this be done? We might wish to
represent the fact that a man's wealth began to grow – perhaps
an inventor's first success. We might show him storing sovereigns
in a safe. The first week he puts a sovereign in. Next week he
adds two more sovereigns. Then he adds a pile of three sovereigns.
The earnings of each week form a pile, and each pile contains one
sovereign more than the previous pile.

Very good – in the xth week he saves £x, and the steadily rising
heaps of coin show at a glance the meaning of this fact.

But the plot need not end there. The inventor has a friend who lives by fraud and violence, a gangster. The gangster is determined to prove that it does not pay to be honest. He urges the inventor to come where the big money can be made. Scornfully, each week, he places a pile of coins behind the inventor's. The first week, £1; the second, £2; the third week, £4; the fourth week, £8 – doubling each week. The steady rise of the inventor's earnings sinks into insignificance. He may save £x – the gangster saves £2^{x-1}, each week. (Fig. 6).

These figures not only show that the incomes of the inventor and the gangster are growing. They bring out the significance of the different ways in which they are growing. Here we have the essential idea of a *graph* – that is, a diagram which will show to the eye the meaning of a mathematical expression such as x or 2^{x-1}.

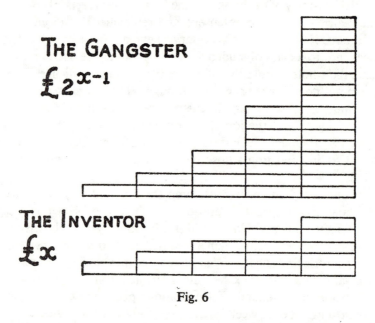

THE GANGSTER
£2^{x-1}

THE INVENTOR
£x

Fig. 6

In this particular illustration we have been considering something which grows by steps, by sudden jumps. For instance, the gangster's weekly savings rise from £4 to £8 without passing through the values between £5, £6 and £7. It is also possible for a thing to grow steadily, like a plant, without jumps. We can, and usually do, draw graphs to illustrate this steady growth. For instance, if a plant grows continuously according to the formula $y = x$ (where y means the height of the plant in inches after x weeks), this not only means that the plant is 1 inch higher after 1 week and 2 inches after 2 weeks. It means that it is $1\frac{1}{4}$ inches high after $1\frac{1}{4}$ weeks, $1\frac{1}{2}$ inches after $1\frac{1}{2}$ weeks, and so on. We show this by a diagram in which the steps have been smoothed out (Fig. 7.).

Whether one draws a continuous curve, or one which rises by steps, is decided by the nature of the process which the graph is intended to illustrate. The size of a plant, the distance travelled by a train, the weight of a child – these give continuous graphs. The number of children in a family, the seats held by a party in Parliament, the number of battleships in a navy – these change by steps.

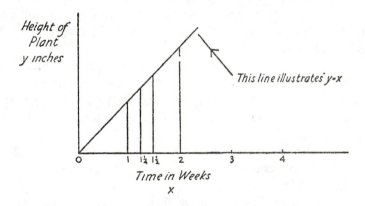

Fig. 7

There are certain cases in which one may use either a continuous curve or a graph with steps. Suppose, for instance, we wish to represent the growth of the population of Great Britain from 1800 to 1900. Strictly speaking, this changes by jumps – increases by one whenever a child is born and decreases by one at every death. But the population itself is measured in millions: to get our graph a reasonable size, we must take a scale such that a million people are represented by not more than an inch. Each individual birth or death corresponds to a change of not more than a millionth of an inch. This is far less than the thickness of a pencil line – even if we could draw each little step, we should be unable to see the effect. So the population curve would appear as a continuous curve – not as a staircase.

Graphs have become so much a part of everyday life that it is not really necessary to explain them. People entirely without mathematical training are usually able to see the significance of the temperature chart over a patient's bed, of the curves showing changes in unemployment or the history of Lancashire's cotton exports. Graphs are used to show the progress of a campaign to raise money, or the output of a factory. Business journals contain graphs showing the trend of prices. Wireless valves are marketed with graphs to show their characteristics. At some holiday resorts one can see instruments which record curves, showing how the barometer has risen and fallen, and charts of the rainfall and sunshine from day to day. The general idea of a graph is already widely understood.

It may be useful to explain exactly how a graph is drawn. A graph illustrates the connexion between two sets of numbers. For instance, we have already considered the possibility that a plant might grow as shown in the following table:

Number of weeks (x)	0	$\frac{1}{4}$	$\frac{1}{2}$	$\frac{3}{4}$	1	$1\frac{1}{4}$	$1\frac{1}{2}$	$1\frac{3}{4}$	2
Height in inches (y)	0	$\frac{1}{4}$	$\frac{1}{2}$	$\frac{3}{4}$	1	$1\frac{1}{4}$	$1\frac{1}{2}$	$1\frac{3}{4}$	2

We drew a level line, along which we marked the number of weeks. We then drew lines upwards, representing the height of the plant corresponding to any number of weeks. In Fig. 7 such lines have been drawn corresponding to 1, $1\frac{1}{4}$, $1\frac{1}{2}$ and 2 weeks. The upright line corresponding to 1 week is 1 inch high – the size of the plant after 1 week. The upright line corresponding to $1\frac{1}{4}$ weeks represents the height of the plant after $1\frac{1}{4}$ weeks. So one can go on, drawing as many upright lines as one likes. These upright lines show the growth of the plant, in the same way that the piles of coins show the growth of the inventor's weekly savings. After drawing a large number of these upright lines, we can see that the tops all lie on a certain straight line. (In other examples the tops lie on a curve.) Drawing the line (or curve) joining the tops of the upright lines, we obtain *the graph of the plant's growth*. As the plant grows according to the formula $y = x$, this line is also called *the graph of* $y = x$.

Any other process, or the mathematical formula which describes it, can be graphed by this method. On page 93 there is a table showing the motion of a ball thrown into the air. Draw for yourself a graph to illustrate this table. The tops of the upright lines will lie not on a straight line, but on a curve. Notice how this curve rises so long as the ball is going up, and descends as the ball comes down. What would the graph of a bouncing ball look like?

In both these examples, y has been a mathematical function of x. For the plant, $y = x$. For the ball, $y = 10x - x^2$. But do not suppose that graphs can be drawn only when a simple formula exists. One can graph the temperature of a patient or the price of milk: it is extremely unlikely that a simple formula can be found to fit either of these.

The Uses of Graphs

Graphs have a great advantage over tables of figures, when information has to be taken in at a glance. It is quite easy to run an eye down a row of figures, and fail to see that one number is much larger than the rest. On a graph, such a number would stand out like a mountain peak. A sudden bend in a graph is easily seen – a casual glance at the corresponding figures would certainly not reveal its existence. Graphs are particularly useful for busy men who want to know the general outlines of a situation but do not want to be bothered by going into every small detail.

This is the simplest use of a graph – to convey a general impression. An historian or an economist may simply want to know that Lancashire was prosperous in 1920 and that a sharp crash

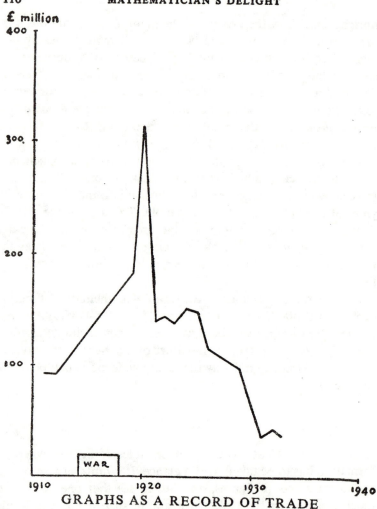

GRAPHS AS A RECORD OF TRADE

The graph shows the exports of cotton cloth, in million pounds, during the years in question. One can see in a few seconds the general outline of Lancashire's fortunes in this period. One would grasp far less from seeing the actual figures. Try for yourself taking a column of figures from an encyclopedia or year-book. Glance at them for fifteen seconds, put them away and write down the things you have noticed – when the figures are largest, when they are small, when they are growing, when they are getting less, etc. You will not notice very much in a short time. Now make a graph of the figures, and notice how the graph brings out things you have missed.

came in 1921. One glance at a graph of cotton exports will remind him of this fact.

Again, graphs can be used to bring out the connexion between two events. Most books on Germany point out how the distress in Germany during the world slump created a mood of extremism and desperation, and helped the rise of the Nazi party. How far can we accept this view as being true? Let us draw, on the same piece of paper, two graphs, one showing the amount of unemployment in Germany, the other showing the number of Nazi M.P.s, between 1926 and 1933 (Fig. 8).

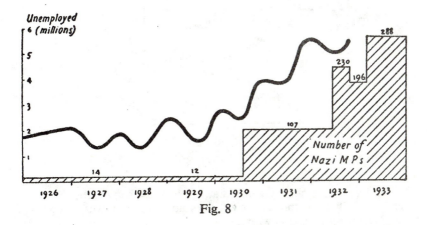

Fig. 8

This graph immediately shows that there is some truth in the idea. During the boom years, the elections returned negligible numbers of Nazi M.P.s: 14 and 12. The two curves, in the main, rise together.

It would, however, be absurd to try to find a mathematical formula connecting the two things. The number of M.P.s changes by steps, at every general election. Unemployment and insecurity are not the only causes acting. For instance, the setback to the Nazis in the second election of 1932 was due to political causes – quarrels among the Nazis, a belief that the Army would oppose Hitler, and so forth.

You will notice how a graph calls your attention to unexplained

GRAPHS USED TO DETECT UNDER-NOURISHMENT

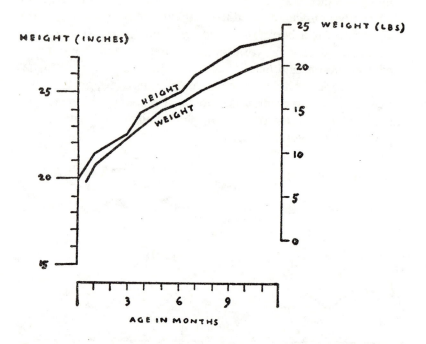

The two graphs show the height and weight of a baby in the first year of its life. If the graph for weight does not keep pace with the graph for height, something is wrong.

facts, and urges you on to further enquiry. We drew the graph to see how far the slump explained the rise of Hitler. The graph not only gives us an indication of the probable answer to this question: it brings to our notice the fall in the Nazi vote at the end of 1932, which is in no way connected with the curve for unemployment, and sets us searching for further facts to explain the setback.

Again, one can hardly help noticing the wave-like shape of the unemployment curve, which rises every winter and falls every summer. This reminds us that there are some trades – such as building – which cease work during bad weather. The summer of 1926 seems to have been an exception. One wonders why. The longer one looks at a graph, the more questions it suggests and the more information it helps one to remember.

It is worth while to collect graphs on any subject in which one is interested. One often hears remarks, and wonders what evidence there can be for their truth. A visit to a public library will often establish the truth or expose the falsehood of a statement. If it is possible to illustrate the question by means of a graph, one has a way of recording much information in a little space. One need not look up the same facts again. As time passes, the collection of graphs is likely to contain some interesting facts.

Quiller-Couch, in *The Art of Reading*, mentions a girl who kept a graph of the attendance at a village church, and tried to account for every rise and fall that occurred. She must have gained an amazing knowledge of the qualities of the preachers and the habits of the village.

Graphs are used by doctors, to show whether children are being properly nourished. The weight of a child, and its height, are graphed on the same piece of paper. For a healthy child, the two curves go up together. If the child is not getting the food it needs, the curve for weight fails to keep pace with that for height. The doctor need not wait until there is a big gap between the two curves. If he notices that the curve for weight begins to bend downwards, this may be the first sign that something is going wrong. If, after special treatment or extra food, the curve begins to bend upwards again, the doctor knows that good effects are beginning to be felt. Part of the science of interpreting graphs consists in knowing how a graph looks when something is increasing, when it is increasing very fast, when it is increasing faster and faster, when it is increasing but increasing slower and slower. (Draw graphs, to illustrate these different possibilities.)

In all these examples the conclusions drawn from the graph have been of a rather general nature. The doctor sees that a child is getting healthier or less healthy, but he does not attempt to *measure* how healthy it is. He cannot say that it is 80% healthy, any more than we can say that someone is 80% happy or 80% honest. Such things as health, happiness and honesty can be measured only indirectly: statistics of deaths, suicides, thefts, may throw some light on these matters. But it is quite possible to know a lot about how healthy, happy or honest a person is, without being able to give a single figure of anything that can be measured.

There are some departments of life, on the other hand, in which measurement plays a large part. This is particularly true of such subjects as engineering, chemistry, physics. Quite a small bump on a railway track may be sufficient to derail a train: if a ball-bearing is one-thousandth of an inch too large, it may take all the weight that should be spread over several ball-bearings and wear out too fast. In such matters very exact calculations are frequently necessary. For this reason engineers and scientists cannot be content with rough statements. They sometimes want to say, not merely that a curve rises slowly, but that it rises at a rate of 1 in 100, or 1 in 87. Much of mathematics has developed in an attempt to satisfy such demands of engineers: mathematicians have been led to invent a whole set of numbers, by means of which one can not only describe, but measure, exactly what a curve is doing at any point. The next chapter – on the study of speed – explains how this is done.

Mathematicians and Graphs

Mathematicians use graphs for many different purposes, some of which are indicated in the following paragraphs.

Graphs may be used to help us to know what we are talking about. It often happens when long calculations are being made with algebraic symbols that we lose sight of the meaning of these symbols; we have at the end a formula, which has been obtained by using the rules of algebra, but we have no way of feeling what it means. It deepens our understanding of the subject if we do not rest content with having found a correct formula, but try to realize what this formula means.

For instance, the formula –

$$P = 364\ V - \frac{V^3}{270,000}$$

gives the power (P) transmitted when a shaft is driven by means of a leather belt. V represents the velocity in feet per second at which the leather belt is travelling. The formula holds in certain conditions which do not concern us at the moment.

What does this formula mean? It contains a quite striking result. It would be natural to suppose that, by turning the driving pulley sufficiently fast, one could transmit as much power as one wished. But draw the graph of P, taking V for values between 0 and 8,000. You will find that P rises until V is 5,700, after which it decreases. If you drive the belt faster than 5,700 feet a second, you do not transmit more power but *less*. One glance at the graph shows this. If, however, one did not draw the graph, but used the formula blindly, one might make serious mistakes, such as designing a plant which worked at a speed too high to be economical.*

Graphs can be very helpful to anyone learning mathematics. Many people can follow all the steps in the solution of a problem, when the solution is shown to them, but they are unable to discover the solution for themselves. They understand each separate step, but they do not know which series of steps will bring them out of the wood. This difficulty can be overcome only if one learns to *see* the meaning of mathematical formulae. Many mathematicians think about their problems all day, wherever they may be. They do not remember all the formulae: they remember a *picture* which the problem has created in their minds. They keep thinking about this picture, until a *method* of solving the problem occurs to them. Then they go home to their pens and paper and collections of formulae, and work out the solution in full. Graphs are one of the ways by which it is possible to form a picture of a problem.

It is a good practice to collect, and to become familiar with, the graphs of the more common functions, such as $y=x$, $y=2x+1$. $y = 3 - 2x$, $y = x^2$, $y = x^2 + 2x + 5$, $y = x^3$, $y = x^4$, $y = 2^x$, $y = \frac{1}{2}x$, and so on.

In scientific work one often obtains a set of results by

* Both the formula and the graph can be found in J. Goodman, *Mechanics Applied to Engineering*, Vol. 1, page 355, 9th edition.

experiment, and then tries to find a formula to fit these results. This can be very difficult, since there are many different types of formula, any one of which *might* be the correct one. It is often helpful to represent the experimental results by means of a graph. If one is familiar with the graphs of many functions, one may at once recognize the type of function which produces such a graph. For instance, all functions which have straight lines as graphs are of the type $y = ax + b$.

Of course, small errors always creep in, and one does not expect the points to lie exactly on a smooth curve. Such small errors in measurement are due to various causes – the thickness of the lines on a ruler when a length is being measured, for instance. Occasionally a big mistake occurs – for instance, one might copy 7197 as 7917, or forget to close a switch when an experiment was being done. Such big mistakes are easily detected on a graph. All the other readings cluster around a smooth curve, but the big mistake is far away from the curve, and one suspects it immediately.

This way of detecting errors is useful not for scientific work only, but also for mathematics itself. For instance, in calculating a set of numbers we may make slips in one or two of the numbers. By drawing a graph it is usually possible to see which numbers are incorrect. When all the numbers are correct, the graph will be a smooth curve – at any rate, this is true in the great majority of cases.

Graphs not only enable us to express a formula by a curve: they enable us to describe a curve by a formula. For instance, when there is no wind, a jet of water from a hose or a small pipe forms a simple curve. If a board is held near the jet of water, the curve can be traced. One can then study this curve, and try to find the formula of which it is the graph. The formula, once found, provides a sort of name for the curve. The part of mathematics known as *Analytical Geometry* is based on this idea of describing every line or curve by the formula corresponding to it. If you want to learn analytical geometry, but find the text-books difficult, the best thing to do is to experiment with graphs for yourself. Draw graphs of the type $y = ax + b$, taking all sorts of values for a and b, positive and negative, large and small. Verify the statement made earlier that all these graphs are straight lines. What do you notice about the graphs of $y = x$ and $y = x + 1$? Can you find a formula which gives a straight line at right angles to $y = x$? Experiment on these lines, record your experiments, and try to reach general conclusions: see how long it is before you can tell, simply by looking at the formulae, that two lines are at right angles. *Then* read the chapter in the text-book headed 'The Straight Line' or 'The Equation of the Straight Line', and you will find your own results, in another person's language. Since you already know what the author is trying to say, it will not be long before you come to understand his language.

EXAMPLES

1. Draw the following graphs. What do you notice about them? How would you describe in words the figure they form?

(i) $y=2x$. (ii) $y=2x+1$. (iii) $y=2x+2$. (iv) $y=5-\frac{1}{2}x$.

2. Draw and describe, as in question 1, the four graphs following.

(i) $y=3x$. (ii) $y=3x+1$. (iii) $y=3x+2$. (iv) $y=4-\frac{1}{3}x$.

3. What do you notice about the following two graphs?

(i) $y=x^2+2x$. (ii) $y=x^2+4x+3$.

4. Draw the graph $y=x(9-x)$. For what value of x is y largest, and what is the largest value of y?

5. What do you notice about the graphs following? –

(i) $y=25-x^2$. (ii) $y=x^2$?

Minus Numbers on Graphs

Often we want to draw a graph, in some part of which x or y or both are minus numbers. For instance, we may want to draw a graph showing the length of an iron rail for temperatures which are *below zero*. If x degrees is the temperature, this means that x is a minus number. If in our graph $x=1$ is one inch to the right, $x= -1$ will be one inch to the left. $x= -2$ will be two inches to the left, and so on.

In the same way, if $y=1$ is one inch up, $y=-1$ will be one inch down.

For instance, we may draw the graph of $y=x-1$, for values of x lying between -3 and 3. We first make the table –

x	-3	-2	-1	0	1	2	3
y	-4	-3	-2	-1	0	1	2

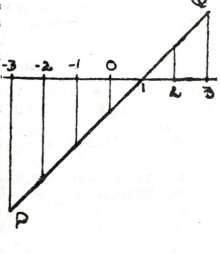

We mark the values of x on a level line. $x=3$ is marked 3 inches to the right of 0, -3 three inches to the left of 0, and so on. The corresponding values of y are then shown as upright lines. When $x=3$, $y=2$, so that a line 2 inches high is drawn from the point where $x=3$ is marked. When $x=-3$, $y=-4$, so a line 4 inches *downwards* is drawn from the point where $x=-3$ is marked. The ends of these lines give us the line PQ, which is the required graph of $y=x-1$.

One of the advantages of using minus numbers will be seen in the examples on the shape of bridges. Very often the formula is much simpler if we take $x=0$ at the middle of the bridge, than if we take it at the end.

6. Draw the graph of $y=x-2$ for values of x between 2 and 6. This gives part of a straight line. With a ruler extend this straight line in the 'south-westerly' direction. Check that this line now passes through the points given by the table for x between -4 and 2.

7. Draw the graph $y=5-x$ for x between 0 and 5. This gives part of a straight line. Extend this line, using a ruler. Read off the

values of y corresponding to $x=6$ and $x=7$. For what values of x is y equal to 6 and 7? Does this agree with the way of finding $5-(-1)$ and $5-(-2)$ explained in Chapter 5?

8. *Graphs to Describe Shapes.* – The book *Building with Steel* by R. B. Way and N. D. Green contains sketches of various famous bridges. The curves there shown seem to agree with the graphs given below –

(i) The Langwies Viaduct, Switzerland. The central arch, of reinforced concrete, resembles the curve –

$$y=2-\frac{2x^2}{9}.$$

(ii) The long low arch of the Royal Tweed Bridge at Berwick,

$$y=1-\frac{2x^2}{37}, \text{ from } x=-4\cdot3 \text{ to } x=4\cdot3.$$

(iii) The lower cable on the suspension bridge at the side of Tower Bridge,

$$y=\frac{9x^2}{80}.$$

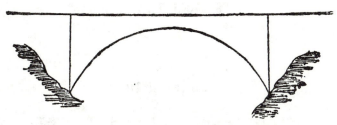

The Outline of Victoria Falls Bridge

(iv) The arch of Victoria Falls bridge,

$$y=\frac{116-21x^2}{120}.$$

The upright lines are to be marked at $x=2\cdot35$ and $x=-2\cdot35$. The level line is at the height $y=1\cdot25$.

9. In each of the following sets of numbers a mistake occurs. On a graph, all the numbers of each set ought to give a smooth curve. Which numbers are incorrect? What are the correct numbers that ought to be in place of the wrong ones? (Only a *rough* answer is expected to the second question.)

(i) 10, 13, 61, 19, 22.
(ii) 4, 11, 13, 19, 20, 19, 16.
(iii) 0, 0, 5, 9, 12, 14, 15, 15.
(iv) 23, 34, 41, 49, 50, 52, 53.
(v) 3610, 4000, 4410, 4640, 5290, 5760.

10. One of the following numbers is wrong, but not much. Can you find which one it is?

3844, 3969, 4096, 4252, 4356, 4489, 4624, 4761.

(Probably you will not be able to do this by the method of question 9. The point of this question is to show what *cannot* be done easily by graphs. The point of question 9 is to show what *can* be done with graphs. Somewhere in this book you will find a method that supplies a clue to the question just asked.)

11. Even if you answer question 10 by graphs, you will certainly need another method to find the wrong number in the set below –

6724, 6889, 7065, 7225, 7396, 7569.

CHAPTER 10

DIFFERENTIAL CALCULUS – THE STUDY OF SPEED

' *When I was an apprentice, I wanted to know engineering theory, and the only two books available contained unknown mathematical symbols . . . Other boys might sigh for luxuries, but to me the one thing wanting was a knowledge of $\dfrac{dy}{dx}$ and $\int$. Looking back, I seem* *to have panted for a knowledge of the use of these symbols for years.'* – John Perry.

ONE of the commonest words in modern life is 'speed'. It is therefore natural that mathematicians, who have had a hand in most of the scientific and industrial advances on which modern life is based, should have a special set of symbols to describe speed, and a special subject dealing with the use of these symbols. Like the sellers of patent medicine, mathematicians have not been able to resist the lure of high-sounding titles. The subject is known by the name of the Differential Calculus.

In any subject dealing with things that move, or grow, or change, you are likely to find the symbols of differential calculus. Even in subjects where nothing seems to be moving, these symbols turn up. We say that a road bends 'very suddenly'; we can discuss how 'quickly' the direction of a railway line changes. Neither the road nor the rail is moving at all. Yet we do mean something when we use such phrases. Words originally meant to describe motion – 'quickly', 'suddenly' – can be used to describe motionless objects. It is the same with the symbols which take the place of words in mathematical work. They too can be used to describe the curve of a road, or a railway, or of any similar object.

Differential calculus is therefore a subject which can be applied to anything which moves, or has a shape, or changes – and this does not leave much out! It is useful for the study of machinery of all kinds, for electric lighting and wireless, for economics and for life insurance. For two hundred years after the discovery of the differential calculus, the main advance of mathematics lay in applications of it. Very few really new ideas came into mathematics. Once the basic ideas of differential calculus have been grasped, a whole world of problems can be tackled without great difficulty. It is a subject well worth learning.

The Basic Problem

The basic problem of differential calculus is the following: we are given a rule for finding where an object is at any time, and are asked to find out how fast it is moving.

For instance, we might be given the following table, for a stone rolling down a hillside.

TABLE VII

Time in seconds	0	1	2	3	4	5	6
Distance gone (feet)	0	1	4	9	16	25	36

This rule, of course, is a very simple one: $y=x^2$, x meaning the time in seconds required to go y feet.

We are now asked: You know where the stone is at any time, find out how fast it is going. Let us try to find out how fast it is going after one second.

First of all, it is easy to see that the stone continually goes faster and faster. In the first second it goes only 1 foot; in the next 3 feet; in the third second 5 feet, and so on, 2 more feet for every second that passes.

But this does not tell us how fast it is going after 1 second, though it helps us to get an idea of the answer. In the first second the stone goes 1 foot. It is *averaging* a speed of 1 foot a second, during this second. This does not mean that its speed is 1 foot a second. A car which travels 30 miles in an hour does not travel at a speed of 30 miles an hour. If its owner lives in a big town, the car travels slowly while it is getting out of the town, and makes up for it by doing 50 on the arterial road in the country. The rolling stone is doing the same sort of thing – it starts slowly, but (so to speak) keeps its foot on the accelerator the whole time. As it covers 1 foot in the first second, its speed at the end of that second must be *more* than 1 foot a second, for it reaches its highest speed (for the first second) right at the end.

Its speed still goes on increasing during the second interval, in which it covers 3 feet. Accordingly, at the beginning of the second interval its speed must have been *less* than 3 feet a second.

Accordingly, after 1 second its speed is somewhere between 1 foot a second and 3 feet a second.

This is the best we can do, if we merely consider the figures for whole seconds. But there is no need to keep to whole seconds.

Our rule, $y=x^2$, applies equally well to fractions. If we work out the distances corresponding to 0·9 second and 1·1 seconds, we have the little table:

TABLE VIII

x	0·9	1	1·1
y	0·81	1	1·21

Just the same argument can now be applied again. In the tenth of a second between 0·9 and 1 the stone covers 0·19 of a foot. This represents an average speed of $\dfrac{0·19}{0·1}$ feet a second – that is, 1·9 feet a second. In the same way, the average speed in the tenth of a second just after 1 second is 2·1 feet a second. The number we want therefore lies between 1·9 and 2·1. In getting this result, we have not had to carry out any process more complicated than ordinary division.

But there is no limit to the accuracy we can obtain by this method. If we consider the hundredth of a second just before, and the hundredth just after, 1 second, we find that the speed lies between 1·99 and 2·01 feet a second. If we take a thousandth of a second, we find the speed lies between 1·999 and 2·001. And there is nothing to stop us considering a millionth or a billionth of a second if we want to. Only one speed will satisfy all these conditions – exactly 2 feet a second. And that is the answer to our question.

In exactly the same way you find the speed after 2 seconds. The little table may then be read:

TABLE IX

x	1·9	2	2·1
y	3·61	4	4·41

which shows that the speed after 2 seconds lies between 3·9 and 4·1. In fact, the speed is 4 feet a second.

So one may work out the speed after any time. The results of doing this are collected in the following table.

TABLE X

Time in seconds	0	1	2	3	4	5	6
Speed (in feet a second)	0	2	4	6	8	10	12

From this table it is easy to see the rule. After x seconds the speed is given by the number $2x$.

By this 'experimental' method it is fairly easy to find the rules for other formulae. First, one finds the speeds corresponding to 1, 2, 3, 4, 5, 6, etc.; then (by the method explained in Chapter 8) one tries to find a formula which will fit this set of numbers, and give the rule for the speed after x seconds. This enables one, at any rate, to *guess* the answer: to *prove* its correctness one has to use algebra.

You should be able to find for yourself the speeds corresponding to the formula $y=x^3$, the speeds corresponding to the formula $y=x^4$, and so on, with $y=x^5$, $y=x^6$, etc. When you have done $y=x^3$ and $y=x^4$, you will notice how simple the answers are. This suggests that the answers for $y=x^5$ and $y=x^6$ will also be simple, and helps you to shorten the work of guessing the rule. To work out the rule by the method of Chapter 8 would otherwise be a rather long piece of work. If you possibly can, work out for yourself the answers for the above cases, without looking at the results below. Anyone can do this work, and it makes all the difference to morale if you can find the results for yourself without ever looking at a text-book.

If you succeed in doing this, you will obtain the results set out in the table below.

TABLE XI

Formula for Distance Gone in x seconds.	Formula for Speed after x seconds.
$y=x^2$	$2x$
$y=x^3$	$3x^2$
$y=x^4$	$4x^3$
$y=x^5$	$5x^4$
$y=x^6$	$6x^5$

It is obvious that these results could be obtained by a simple rule. Where we have x^3 in the first column, we have something containing x^2 in the second; opposite x^4 stands a certain number of times x^3. The power of x in the second column is always one less than in the first. Opposite x^n there will be something containing x^{n-1}. Even simpler is the rule for the number which stands before x: it is the same as the number in the first column: it is n. Our rule is, 'If the formula for the distance is x^n, the formula for the speed is nx^{n-1}.'

Note what a lot of work is contained in this little result. To find the speed corresponding to x^2 we had to do a whole row of calculations; then we had to notice that the formula $2x$ would fit the results. We had to do this work also for x^3, x^4, x^5 and x^6. We then collected the formulae together in Table XI, and noticed that they could all be fused in one general rule.

Once having found the general rule, we can apply it immediately to any other case. To the formula $y=x^{17}$ will correspond the speed $17x^{16}$, the speed corresponding to x^{92} is $92x^{91}$.

In mechanics, and other applications, we often have to deal with formulae which contain several powers of x. For instance, if a ball is thrown straight upwards with a speed of 40 feet a second, its height after x seconds is given by $40x-16x^2$ feet. (We had this formula in Chapter 8 in a slightly different form. There it was given for the height after x *quarter*-seconds.) How is the speed after x seconds to be found?

The best way to deal with such a question is to split it up. We shall consider in turn the different parts which make up the problem.

(i) How quickly does the term $40x$ grow? $40x$ is the distance that a body would go in x seconds if it travelled with a *steady* speed of 40 feet a second. So the speed corresponding to $40x$ is obviously 40.

(ii) How quickly does the term $16x^2$ grow? We could get the table for $16x^2$ by multiplying all the numbers in the lower row of Table VII by 16. In other words, if a body travels according to the formula $16x^2$ after any number of seconds, it will have gone 16 times as far as one travelling according to the formula x^2. At any moment it must therefore be travelling 16 times as fast. But the speed corresponding to x^2 is $2x$. The speed corresponding to $16x^2$ must be 16 times as large: it must be $32x$.

(iii) We now know that $40x$ grows steadily at the rate 40, while $16x^2$ grows at the rate $32x$. How fast will $40x$—$16x^2$ grow? How are the two rates to be combined?

$40x$—$16x^2$ is obtained by subtracting $16x^2$ from $40x$. How can we picture this subtraction? We might think of $40x$ as representing a man's income at any moment, $16x^2$ as representing the expense of bringing up his family. Both income and expenditure are growing. $40x$—$16x^2$ represents the weekly balance which the man has, after meeting his expenses. It is obvious that this balance will increase at a rate given by the rate at which his income rises, *minus* the rate at which his expenses rise. (If the balance is decreasing, this rate will be less than nothing – it will have a minus sign.) The rate of increase of the income is 40; the rate of increase of expenditure is $32x$. The rate of increase of the balance is therefore 40—$32x$. So we combine the rates by subtracting the second from the first.

We thus reach our grand conclusion: the speed corresponding to the formula $40x$—$16x^2$ is 40—$32x$.

You will see that the arguments used could equally well be applied to any other expression of a similar type. For instance, the speed corresponding to $4x^3+x^2+3x+1$ is $12x^2+2x+3$. (The number 1 does not change at all: $y=1$ would mean that a body always stayed at a distance of 1 from some fixed point. Its speed would of course be nothing at all. So this 1 in the formula does not add anything to the answer. $4x^3+x^2+3x$ would lead to the same speed. This is quite reasonable. $4x^3+x^2+3x$ is always 1 less than $4x^3+x^2+3x+1$. The first formula neither catches up nor falls behind the second. So the speeds are naturally the same).

If you have any difficulty with this idea, reason out for yourself the speeds which correspond to $5x^2$ and to $2x$; then to $5x^2+2x$. Work out the speeds which correspond to x^2+x, x^2—x, $x+1$, x^2+x+1, and other examples which you can make up for yourself. Check your answers by working out the tables for these formulae, and seeing if your answers for the speeds are reasonable.

Signs for Speed

It is inconvenient to keep on saying 'the speed corresponding to the formula'. A sign is therefore used. If we have any formula giving y, the corresponding speed is represented by y'. This enables us to state a rule we had earlier in the shorter form, 'If $y=x^n$, $y'=nx^{n-1}$.' This means exactly the same as 'To the formula x^n corresponds the speed nx^{n-1}.' Just as $y(x)$ is used to represent the distance corresponding to x, so $y'(x)$ is used to represent the speed corresponding to x. Thus $y'(2)$ will mean the speed after 2 seconds.

It is sometimes convenient to use another sign, instead of y'. This other sign is $\dfrac{dy}{dx}$.

There is a reason why this sign is used. The d in it has a very special meaning, like the Δ used in Chapter 8. In fact, it is only through the sign Δ that you can see why d is used here.

What, after all, is a speed? If you are told that a train has covered 300 miles in 4 hours, you know that its speed has been (on the average) 75 miles an hour. The 75 is found by dividing 300 by 4. If you know that a train has travelled 150 miles by 7 a.m. and 270 miles by 10 a.m., how do you find its average speed between 7 a.m. and 10 a.m.? You find the time that has passed between 7 a.m. and 10 a.m., 3 hours. You find the change in the distance, $270-150=120$. Then, dividing 120 by 3, you have your answer: 40 m.p.h.

If we call the time x hours, and the distance gone y miles, we have a table, rather like those in Chapter 8.

TABLE XII

Δx		3	
x	7		10
y	150		270
Δy		120	

As before, we have the values of x in one row, the corresponding values of y beneath them, and then a row giving Δy, the change in y. The only new feature is the row labelled Δx, giving the change in x. In Chapter 8, the change in x, between any number and the next, was always 1, so it would have been a waste of time, as well as a complication, to have had a row Δx. But for finding speeds Δx is absolutely essential. The speed of 40 m.p.h. was found by dividing 120 by 3: that is, by dividing Δy by Δx.

The rule for finding *average speed*, then, is to work out the change in distance divided by the change in time. In our symbols,

$$\text{average speed} = \frac{\Delta y}{\Delta x}.$$

But this only gives *average* speed. We are looking for the speed at any moment. If a car runs into you at 60 m.p.h. it is no comfort to your widow to be told that it had averaged only 10 miles over the last hour, because the driver had spent most of the hour in a pub. The thing that matters is not the average speed during the past hour: it is the actual speed at the exact moment when the car hits you that counts.

But the speed at the moment of the collision will not differ very much from the average speed during the previous tenth of a second. It will differ even less from the average speed for the previous thousandth of a second. In other words, if we take the average speed for smaller and smaller lengths of time, we shall get nearer and nearer – as near as we like – to the true speed. For most practical purposes, the average speed during a thousandth of a second may be regarded as the exact speed.

It is for this reason that mathematicians find it useful to represent *speed* by a sign similar to that for *average speed*. The sign Δ is used by the Greeks to represent the capital letter D. We cannot take over the sign $\frac{\Delta y}{\Delta x}$, unchanged, to represent the speed: for the average speed over a short interval, however near it may be to the exact speed at any moment, is never quite the same: it would lead to confusion to have the same sign for two different things. But, as it were for old times' sake, to remind us how the idea of average speed has helped us towards finding the exact speed, we replace the Greek Δ by an English d, and use $\dfrac{dy}{dx}$ to represent the speed. For the moment, do not ask what dy and dx mean separately. Just regard $\dfrac{dy}{dx}$ as a sign that can be used instead of y', to represent the speed.

Again, in mechanical problems the speed is usually called by the technical term 'velocity', and we may use v as an abbreviation.

The process of finding how quickly a quantity changes is known as *differentiation*. If we differentiate y, we obtain its *rate of change* (or *speed*), y', $\dfrac{dy}{dx}$, v. If we differentiate x^2 we obtain $2x$.

This process can be repeated. When we considered a stone rolling downhill, according to the formula $y = x^2$, we saw that the velocity v continually increased. One might easily ask, 'How fast does it increase?' There is no difficulty in answering this question. We have seen that the velocity v after x seconds is given by the formula $v = 2x$. Thus we have a very simple formula for v, and it is easy to find v'. In fact, $v' = 2$. The velocity increases steadily; it increases by 2 for every second that passes. (Check this result from the values of v given in Table X.)

Since v is the same thing as y', it is natural to represent v' by y''. There is nothing new involved in this sign y''. y' represents the rate at which y changes: y'' represents the rate at which y' changes. In Chapter 8 we found $\Delta^2 y$ from Δy by just repeating the process we had already done to find Δy from y. It is the same here. We start with y. How quickly does y increase? The answer is y'. We now start again with a table (or a formula) for y'. How quickly does y' increase? The answer is y''. In many ways y'' resembles $\Delta^2 y$.

The Importance of y' and y"

The quantities y' and y'' have great importance in mechanics. It is obvious that y' (which simply means the speed, or velocity) is important. If anything, y'' is even more important. y'' measures how quickly the speed changes. If you are in a car travelling at 50 m.p.h. and the driver gradually brings the car to rest – say, within 10 minutes – you feel hardly anything at all. If, on the other hand, the car is brought to rest in one-hundredth of a second – by colliding with a stone wall – this is felt as a blow of tremendous force, sufficient to do serious damage. It does not hurt to travel at a high speed, such as 50 m.p.h. What does hurt is *a sudden change of speed.*

Usually, when our speed changes we feel pressure. If you are in a car, and the brakes are suddenly put on, you feel yourself thrown forward. What really happens is that you keep on moving at the same speed, but the car stops. You stop only when you strike the seat in front of you: you feel this pressing you back. In the same way, a bicycle cannot be slowed down if it has no brakes (unless perhaps there is a strong wind blowing against it, or it is going uphill, or it is badly oiled – any of these states can take the place of brakes).

Newton's Laws of Motion express this idea. According to Newton, if a body could get right away from all outside influences – far away from the pull of the earth and the sun, not being pressed or pulled by any other body, away from electric and magnetic forces – that body would keep on moving in a straight line at a steady speed. With a telescope one can observe little bits of matter, such as comets, and it is seen that the farther away they get from the earth and the sun, the more nearly they move in straight paths at a steady speed.

Whenever we find a body moving in a curved path, or with a changing speed, we therefore believe that something else is interfering with it, is acting upon it. We say that a *force* is acting on it and we try to discover what this force is. Is the body tied to a rope or string? Is it nailed to some other body? Is it being pulled by the earth or the sun or (in the case of tides) the moon? Has it been magnetized? Is it charged with electricity? Is it sliding on a rough surface, which is causing it to slow down? Is it moving in some liquid, such as water or treacle, which opposes its motion? Is it passing through the air, like a parachute, or a falling feather?

Next we ask how to measure the force. It is clear that the force needed to alter the speed of a body depends on how massive the body is. It is easy to stop a runaway pram; harder to stop a runaway wagon; very hard to stop a string of loaded trucks; almost impossible to stop an avalanche. Scientists therefore use the word *mass* to express this quality of a body. One cubic centimetre of water is chosen as the unit of mass; it is called a gramme. Anything which is just as easy to stop, or to get going, as a cubic centimetre of water is said to have a mass of 1 gramme. Anything which is as hard to stop as 100,000 cubic centimetres of water is said to have a mass of 100,000 grammes. And so on. For short, we shall call the mass of a body m grammes.

It is then found that the force which changes the speed of a body (mass m grammes) at the rate y'' is given by my''. When the body is going faster and faster, with y increasing, the force pulls the body forward. When the body is slowing down, y'' is negative: this means that the force is acting as a brake – it is pulling the body back.

In scientific work it is usual to measure the distance y, not in feet or inches, but in centimetres. By using the decimal system of measures we save all the complications of having 12 inches one foot, 3 feet one yard, 22 yards one chain, $30\frac{1}{4}$ square yards one pole, and so on. The English system of measurement has been handed down from very ancient times, long before science or modern engineering had been thought of. It is based on such things as the amount of land a team of oxen can plough in a day (furlong=a furrow long, etc.) or the average size of a part of the human body (*e.g.*, a foot). Such measures were convenient for their original uses. The French system, on the other hand, was introduced during the French Revolution of 1789, and was specially designed for modern trade and industry. Any difficulties that are found in changing from feet and tons to centimetres and grammes must be regarded as due to human history: they are not purely scientific problems.

English engineers also use a system of measurement, in which the standard mass is a one-pound weight, and the distance y is measured in feet. The force corresponding to m pounds and a speeding-up y'' (measured in feet and seconds) is then my'' *poundals*.

We have already considered the formula $y=40x-16x^2$, which gives the height in feet of a body x seconds after it has been thrown-up with a speed of 40 feet a second. What force is acting on this body? $y'=40-32x$. $y''=-32$. If the body has a mass m pounds, the force acting on it is my'', which equals $-32m$. This force does not depend on x. It is the same whatever x may be. The size of the force is $32m$: the sign in front of it is *minus* because the earth drags the weight *down*, as everyone knows. In the case of something which tended to rise (such as a balloon) the force would be $+$.

It is found by experiment that any heavy body thrown into the air moves in such a way that $y''=-32$. We suppose the body is sufficiently heavy for air resistance to be neglected. This law obviously does not hold for a feather, or for a parachute. The whole point of a parachute is that it falls in a way very different from a brick. The law just stated gives good results for a falling stone, cricket ball, or man. It does not work for feathers, rain-drops, or mice. Nor does it work for very high speeds. In the motion of a shell or a bullet, the force due to air resistance may well be greater than that due to gravity.

The figure 32 is of course not exact. The earth does not bother to pull us towards itself with a force that is a simple multiple of the length of our feet! But 32 is near enough for most purposes.

Since the pull of the earth causes y'' to have the value -32, the force exerted on a mass of m pounds by the earth must be $-32m$ poundals (found by putting $y''=-32$ in the expression my''). The *minus* sign means that this force acts *downwards*.

*Other Useful Subjects**

So far we have dealt with a very special case, that of a body moving in a straight line, and acted on by a single force only.

* The remainder of this chapter contains applications which may interest some readers, but are not necessary for an understanding of the rest of the book. A brief reference to this section is made in Chapter 13.

In nearly all practical studies the problem is more complicated. A lift moves up and down in a straight line, but two important forces act on it – the pull of the earth downwards, and the pull of the supporting rope upwards. We may also need to take into account any device used to stop the lift bumping against the walls of the shaft, friction, air-resistance, etc. – even if we neglect these, we still have *two* forces to consider. In other examples we may have to deal with bodies that do not move in straight lines: a train or motor-car going round a bend, a shell in the air, a piece of metal in a flywheel.

Statics deals with the combined effect of a number of forces. Its laws have to be discovered in the first place by experiment, and can then be used for reasoning.

When several forces act in the same direction, the result is what you might expect. If two men pull a sledge, each hauling with a force of 1000 poundals, the effect is the same as that of a single pull of 2000 poundals: the separate forces are simply added together.

When two forces act in opposite directions, the effect is easy to calculate. If a lift weighs 2500 pounds, the earth will pull it downwards with a force of 32×2500 poundals – i.e., 80,000 poundals. If the wire rope pulls the lift upwards with a force of 100,000 poundals, two forces are acting on the lift, 100,000 poundals upwards, 80,000 poundals downwards. These two forces combined have the same effect as a force of 20,000 poundals (obtained by subtracting 80,000 from 100,000), acting upwards. Note that the rope must be able to stand a strain which is *greater* than the weight of the lift. For it is only when the pull of the rope upwards is greater than the pull of the earth downwards that the total force is upwards. And one must be able to make the total force act upwards in order to start the lift for an upward journey and (equally important) to check it at the end of a downward journey. In a coalmine, the cage lifts and lowers the miners through hundreds of yards in an amazingly short space of time. Very great changes of speed take place, and it is a matter of vital importance that the rope be strong enough, not merely to bear the weight of the cage and the men inside it, but also to exert the extra force necessary for starting and braking. (It is possible to experiment with model lifts, using cotton thread instead of rope, to demonstrate the effect of a sudden jerk.)

An entirely new principle has to be learnt in order to deal with forces which do not act in the same straight line. Suppose we have a force of 2 poundals acting East, and a force of 3 poundals acting North: to what single force is this equivalent? It is impossible to answer this by argument: we can only try to see what happens when a small weight is dragged towards

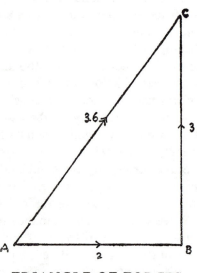

TRIANGLE OF FORCES

the East by one string attached to it, and towards the North by another. (For details of the experiment, see text-books on Statics.) The reader will be able to *feel* what sort of result is likely – the weight will be dragged in a direction somewhere between North and East. Experiments show that the following method gives the correct answer. Draw a line 2 inches long, towards the East. Call this line AB. From the Eastern end of this line (B) draw a line BC, 3 inches long, towards the North. The line AB is drawn 2 inches long, to represent a force of 2 poundals; the line BC is made 3 inches long, corresponding to the force of 3 poundals. If we measure AC, we find it to be 3·6 inches long. The length and direction of AC gives the answer to the question. The two forces, 2 poundals towards the East and 3 poundals towards the North, acting together, will drag the weight in the direction AC, with a force of 3·6 poundals. This principle is known as the Triangle of Forces, for an obvious reason. In the triangle ABC the two sides AB and BC represent the two forces given. The third side, AC, represents the single force produced by these two forces acting together.

In an ordinary catapult two pieces of elastic are fastened to a small piece of cloth. When the catapult is fired, the small piece of cloth moves in a direction that lies between the direction of the two pieces of elastic.

Co-ordinate Geometry

In dealing with graphs we used the idea of fixing the position of a point by measuring the distance across the paper ('to the East') and the distance 'to the North'. The same idea may be used to study the movement of any small weight, when it is not moving in a straight line. We suppose that, after x seconds, the small weight is y feet to the East and z feet to the North of some fixed landmark, O. The small weight might be part of some machine. The rules giving y and z in terms of x will depend on the way in which the machine is constructed. For instance, the weight might be some part of a locomotive. If we know the shape of the rail-way line, and the speed at which the train is travelling, we know *where* each part of the locomotive will be at any time. In other words, we know what values y and z will have after x seconds.

If no force is acting on a body, the body moves in a straight line. When a locomotive goes round a bend, it is not moving in a straight line, nor is any piece of the locomotive moving in a straight line. So there must be forces acting on each part of the locomotive. You will always notice that a locomotive, in going round a bend, presses against the outer rail, just as a motor-car going round a corner too fast tends to run into the outer edge of the road, if the road is not sufficiently banked. The rail presses back on the wheels, and makes these go round the bend, instead of going straight on, as they would prefer to do. Is it possible to find how large is the force acting on any part of the locomotive? It is possible, though it is not easy to describe the method in a few words.

First of all, in order not to make the problem too complicated, let us choose a part of the locomotive which does not move up and down. As it always stays at the same height, its motion can be described completely by giving a table, showing how far it has moved to the East of the landmark O, and how far it has moved to the North of O. Our first task, then, is to discover formulae, or to make tables, giving y and z corresponding to any time, x seconds. We suppose this part of the job completed.

It would be quite easy to study the motion of the locomotive (or the little part of it) towards the East, *if* there were no motion towards the North. The locomotive would then be moving due East, *in a straight line*. After x seconds its distance to the East would be given by y, and the force pushing the small part (of mass m pounds, say) towards the East would be (by our previous method) my''.

It would also be easy to find the answer, if the locomotive was moving due North. The force pushing the small part to the North would be mz'', by a very similar argument.

At this point we receive a free gift from Nature. *It turns out to be true* – and we had no reason to expect this – that the movement towards the East and the movement towards the North can be treated as if they were quite separate. The actual force pushing the small piece is obtained by combining (by the Triangle of Forces) a force my'' to the East and a force mz'' to the North.

The problem therefore can be completely solved. There is no essential difficulty brought in if we consider movements up and down, as well as to the East and to the North. The forces due to the fact that parts of a locomotive move up and down are very important. Old-fashioned locomotives, if driven fast, would jump into the air.

Of the design of modern locomotives, Kempe's *Engineer's Yearbook* writes, 'The horizontal forces are the most injurious, though American engineers consider the vertical forces to be so; but English practice is to take a medium course between excessive horizontal and vertical disturbing influences.'

The calculation of the forces brought into play by moving weights is a practical question in the design and balancing of machinery. In this short space it has not been possible to explain the method in a satisfactory way, but the fact that the method can be outlined, even vaguely, in so few words shows that the principles involved are both few and simple.

Conclusion

Statics and Dynamics will seem very unreal to you if you have had no experience of handling heavy weights. You can learn more dynamics in an afternoon starting and stopping a heavy (but well oiled) railway truck or garden roller than you can from whole books of dynamics. You can get the benefit of reading a book on dynamics only if such words as 'force' call up a vivid image in your mind. Once you have the necessary feel for the subject, the books can be most valuable, even interesting – but not before.

Calculus does not need the same experimental background. Almost everyone already knows what speed is. The job is rather to study sets of figures, until you realize what sort of motion they represent. Take any formula. Work out a table, showing the distance gone after various times. If you cannot see what the exact speed is, begin to ask questions. Silly ones are the best to begin with. Is the speed a million miles an hour? Or one inch a century? Somewhere between these limits. Good. We now know something about the speed. Begin to bring the limits in, and see how close together they can be brought. Study your own methods of thought. How do you know that the speed is less than a million miles an hour? What definite evidence does the table show to support this view? What method, in fact, are you unconsciously using to estimate speed? Can this method be applied to get closer estimates?

You know what speed is. You would not believe a man who claimed to walk at 5 miles an hour, but took 3 hours to walk 6 miles. You have only to apply the same common sense to stones rolling down hillsides, and the calculus is at your command.

EXAMPLES

It has already been urged that the reader should not try to reason about any problem until he has a perfectly clear picture of it in his mind, and has found some way of bringing the problem into touch with real life, so that he can see and handle the things of which it speaks. This is particularly important in the study of *speed*, which is by no means such a simple thing as we at first think it to be. The reader must find for himself some device by means of which he can observe movement. This may be a pencil rolling down a desklid, a bicycle on a hillside, or a weight hanging by a string. One particular device may be mentioned, on the lines of cinema cartoons. Most school-children are familiar with a way of drawing pictures on the leaves of a book, so that when the leaves are allowed to fall in rapid succession the figures seem to move. The same idea may be used to study the movement of a point. It has the advantage that one can study the movement 'frozen' – by looking at the points marked on the various pages of the book – as well as in action. In questions 1 and 2 it is assumed that the leaves of the book fall at the rate of ten each second.

1. On the first sheet of a book mark a point $0 \cdot 1$ inch from the bottom of the page, on the second $0 \cdot 2$ inch, etc., the point on the nth sheet being $\frac{n}{10}$ inches up the page. This illustrates the movement in which $y=x$ (y in inches, x in seconds). That the point moves with a steady speed is shown by the fact that the mark on any sheet is always $0 \cdot 1$ inch higher than that on the page before.

2. On the nth sheet mark a point at the height $\frac{n}{100}$. This corresponds to the movement $y=x^2$ discussed in this chapter. Notice how slowly the point moves in the first half-second (five sheets), how it gains speed as time passes.

3. A body moves according to the law $y=x$. Make a table for its motion and convince yourself: (i) that it is moving at a steady speed, (ii) that this speed is 1. In fact, when $y=x$, $y'=1$.

4. Similarly, show that when $y=2x$, $y'=2$.

5. And when $y=\frac{3}{4}x$, $y'=\frac{3}{4}$.

6. When $y=x+1$, $y'=1$.

7. And when $y=x+2$, $y'=1$.

8. When $y=\frac{3}{4}x+1$, $y'=\frac{3}{4}$.

9. When $y=\frac{3}{4}x+2$, $y'=\frac{3}{4}$.

10. A dog is chasing a cat. The cat moves according to the formula $y=30x+2$, the dog according to the formula $y=30x$. Is it true that (i) both animals are moving at 30 feet a second; (ii) the dog is always 2 feet behind the cat; (iii) $y'=30$ for both formulae?

11. If the cat moves according to $y=20x+10$ and the dog according to $y=25x$, is it true (i) that the dog starts 10 feet behind the cat; (ii) the dog moves faster than the cat; (iii) the dog will catch the cat within a short time? What is y' for the cat? for the dog? When will the dog overtake the cat?

12. Write down y' in the following cases:
(i) $y=x^2$. (ii) $y=2x^2$. (iii) $y=2x^2+1$. (iv) $y=\frac{1}{2}x^2$. (v) $y=x$. (vi) $y=x^2+x$. (vii) $y=x^2+x+1$. (viii) $y=\frac{1}{2}x^2-1$. (ix) $y=x-x^2$. (x) $y=1-x^2$. (xi) $y=x^3$. (xii) $y=2x^3$. (xiii) $y=2x^3+x$. (xiv) $y=2x^3+1$. (xv) $y=10x^3-20x^2+7x-3$.

13. We have seen that when $y=x^2$, $y'=2x$ and $y''=2$. Make tables showing y, y', y'', Δy and $\Delta^2 y$ for $x=0, 1, 2 \ldots 10$. Is it true that (i) the table for y' is rather like, but not quite the same as, the table for Δy; (ii) $\Delta^2 y$ is exactly the same as y''?

14. If $y=x^3$, $y'=3x^2$ and $y''=6x$. Make tables for y, y', y'', Δy and $\Delta^2 y$. Is it true (i) y' behaves rather like Δy; (ii) y'' behaves rather like $\Delta^2 y$?

15. If you had worked out a problem and found a formula for y' which behaved in a way quite different from Δy (for instance, y' getting steadily larger while Δy got steadily smaller), would you think you had made a slip in your work or not? What about y'' and $\Delta^2 y$? Do you expect them, as a rule, to behave in more or less the same way?

CHAPTER 11

FROM SPEED TO CURVES

'Our townsman, Dr Joule . . . instanced the porpoise, with its bluff figurehead, attaining a velocity of over thirteen miles an hour, whilst voracious fishes are so constructed that they can attain a much greater velocity. He advocated a study of natural proportions to those who wish to be successful shipbuilders.' – Bosdin Leech, *History of the Manchester Ship Canal.*

So far we have considered y' or $\dfrac{dy}{dx}$ simply as a sign for the speed of a moving point. For many important applications this is quite sufficient. But it is only one half of the story. There are many problems for which calculus can be used to describe shape. For instance, it is possible to find the curve in which a chain hangs when its ends are held, or the way in which the strain on a bridge is distributed. Movement does not come into either of these questions at all.

It is very easy to translate our discoveries about movement into statements about shape. Any type of movement can easily be

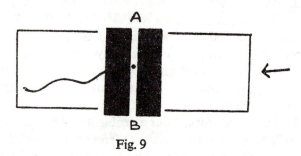

Fig. 9

represented by a curve. Consider the simple arrangement shown in Fig. 9. The point of a pencil is supposed to be moving in the slit AB. Underneath the slit is a sheet of paper, which is made to move at a steady pace towards the left. It is clear that the movement of the pencil will be recorded on the paper as a curve. If we want to see how the pencil was moving, we have only to pass the paper under the slit again. Through the slit we shall be able to see a very short piece of the curve, which will be seen as a point, and will seem to move up and down as the paper passes to the left.

This arrangement is rather similar to a gramophone. The groove of the gramophone record is cut by a vibrating needle. When the record is played, the original vibrations are reproduced. All the peculiarities of the original motion, then, are somehow preserved in the shape of the groove; and any alteration in the shape of the groove would result in some difference when the record was played.

In the same way, the motion of the pencil point and the curve traced on the moving paper are closely connected. Anything that can be said about the movement of the pencil must tell us something about the shape of the curve. Anything that can be said about the curve tells us something about the movement of the pencil.

Now we know that y' tells us how fast the pencil is moving at any moment, and y'' tells us whether the pencil is speeding up or slowing down. It must be possible to find out what y' and y'' represent, what they tell us about the shape of the curve traced by the pencil. To do this is our next job.

The Case of Steady Movement

We will begin by considering the simplest case. Figures 9A to 9E show the traces left on the paper in five experiments. In each of these experiments the pencil moved at a steady speed. The strip of paper is 1 inch wide, and it moved through the slit,

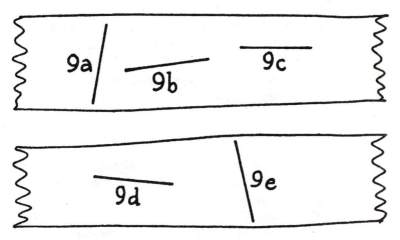

towards the left, at the rate of 1 inch a second. You can, if you like, make an arrangement of the type illustrated in Fig. 9, and pass these strips through it, thus reproducing the original movements.

What happens when 9A passes through? It takes only one-fifth of a second, and during this time the moving point gets right across the paper, a distance of 1 inch. 9A is the track of a point moving at 5 inches a second. For it, $y'=5$.

9B is the track of a point that moves up the slit, but at a much slower rate. In one second the point has risen only one-tenth of an inch. Here $y'=\frac{1}{10}$.

Already, by comparing 9A and 9B, we can see that the line is *very steep* when the point has moved *very fast* (that is, for large y'), but is *nearly level* when the point is moving *slowly* (y' small). In fact, y' *measures how steep the graph is.*

9C is the track left when the pencil-point is *at rest*. The point stays at the same height. It has *no speed*, so $y'=0$.

Accordingly, when $y'=0$, the graph is *level*.

In 9D the pencil-point moves *down* the slit as the paper passes through. When a second has passed, the point has gone down one-tenth of an inch. The change in y during 1 second is therefore $-\frac{1}{10}$, from which it follows that $y'=-\frac{1}{10}$.

Again, in 9E the pencil-point descends 1 inch in one-fifth of a second; it is therefore descending at the rate of 5 inches a second, and $y'=-5$.

We notice that the graph slopes *downhill* when y' is minus (cases D and E) *uphill* when y' is plus (cases A and B).

To sum up: the *steepness* of the graph depends on how *big* y' is: whether the graph goes *uphill* or *downhill* depends on whether y' has a $+$ or a $-$ sign. $y'=0$ means that the graph is *level*.

The General Case

So far we have been considering only what happens when the pencil-point moves at a steady speed. But as a rule this is not the case. We may often have to study things that move with different speeds at different times.

It is, however, still possible to use the conclusions to which we were led by a study of the simpler cases. You should test this for yourself, by means of some arrangement on the lines of Fig. 9. If you move a pencil up and down the slit, varying its speed, you will find that when the pencil is moving fast, the track it leaves is steep; when it is moving slowly, the track it leaves is not very steep. We can still say that *speed* corresponds to *steepness*. If the *speed varies*, then the *steepness* of the graph also *varies*. In this case the graph will be curved, instead of straight, as it was before.

This brings us to the question of y''. y'' tells us how quickly the speed, y', varies. We shall be mainly interested in the sign of y'', whether it is $+$ or $-$. If y'' is $+$, it means that y' is growing (i.e., that y' is changing by having something *added* to it). If y'' is $-$, it means that y' is decreasing (is having something *taken away* from it).

Look at the four sample tracks shown in the figure.

SAMPLE 1.	SAMPLE 2.	SAMPLE 3.	SAMPLE 4.
y' $+$	y' $+$	y' $-$	y' $-$
y'' $+$	y'' $-$	y'' $+$	y'' $-$

What are the signs of y' and y'' in Sample 1? This curve is rising, its slope is uphill; y' therefore must be $+$. The farther you go, the steeper this curve gets. Its steepness (measured by y') is getting bigger. So y' must be increasing. That means y'' must be $+$. It is easy to get a little confused between the meanings of y' and y''. Remember, y' measures how fast y is changing – that is, y' measures the speed of a moving point. y'' measures how quickly y' is changing – that is, how quickly the speed changes.

If this curve were part of a chart showing the course of a military campaign, it would mean (*a*) that the army was advancing, (*b*) that the speed of its advance was continually increasing. (*a*) corresponds to the mathematical statement that y' is $+$; (*b*) to the statement that y'' is $+$.

We have a different state of affairs in Sample 2. True, the slope of the curve is uphill, but the farther you go, the *less* steep it is. This corresponds to the military communiqué 'our advance continues, but is being slowed down by determined resistance'. Since it is an *advance*, y' is $+$. Since the rate of advance is being *slowed down*, y'' is $-$.

One has to be rather careful with Samples 3 and 4, owing to the fact that y' is minus. We have to remember that a change from $y'=-10$ to $y'=-1$ represents an increase in y', owing to the properties of negative numbers.

In Sample 3 the state of affairs at the beginning represents a *rapid descent* – in military terms, a rout. Later on the curve is still doing downhill, but not nearly so fast. The retreat is being checked. To this extent the situation is improving. The *improvement* is reflected by the fact that y'' is $+$. The *retreat* is shown by the downward slope of the curve: y' is $-$.

The reader may remember that the line 9E, going steeply downwards, had $y'=-5$, while 9D had $y'=-\frac{1}{10}$. In Sample 3 the early part of the curve slopes like the line 9E, while the end of the curve is more like 9D. In Sample 3, then, y' begins by being about -5, and ends by being about $-\frac{1}{10}$. You have to add something to -5 to make it into $-\frac{1}{10}$. That is why y'', the rate of change of y', is $+$.

In Sample 4, on the other hand, the situation is going to the dogs faster and faster. The curve is going downhill, becoming steeper and steeper. y' may be about $-\frac{1}{10}$ at the *beginning*, and -5 at the *end*. y' is therefore *changing for the worse* – that is, y'' is $-$.

In these four samples we have covered all the main possibilities. y' must be either $+$ or $-$, and y'' must be $+$ or $-$ (unless y' or y'' should happen to be 0). By combining our four Samples, we can see what the shape of any graph will be, provided we know how y' and y'' behave. Note that it is quite easy to give a simple meaning to y''. When y'' is $+$, the curve *bends upwards* (Samples 1 and 3); when y'' is $-$, the curve *bends downwards* (Samples 2 and 4).

An Example

Suppose you were asked. 'What is the general shape of the graph $y=x^3-3x$?' We will consider the shape of the graph between $x=-20$ and $x=+20$.

First of all we need to know y' and y''. Since $y = x^3 - 3x$, $y' = 3x^2 - 3$, $y'' = 6x$.

Clearly, y'' is $+$ when x is $+$, and $-$ when x is $-$. This tells us that the graph *bends upward* for all positive x, *downward* for all negative x.

If you try a few values for x, you will see that $3x^2 - 3$, the formula for y', is $+$ when x lies between -20 and 1, and also when x lies between 1 and 20. $3x^2-3$ is $-$ when x lies between -1 and 1.

We can collect this information together in a diagram, thus:

x	-20	....	-1	......	0	......	1	....	20
y'		$+ + +$	0	$- - -$		$- - -0$		$+ + +$	
y''		$- - -$	$-$	$- - -$	0	$+ + +$		$+ + +$	
Curve resembles sample		2 2 2		4 4		3 3		1 1 1	

We thus conclude:

From $x = -20$ to -1 the curve looks like SAMPLE 2.

       -1 to 0   „   „   „   SAMPLE 4.

        0 to 1   „   „   „   SAMPLE 3.

        1 to 20   „   „   „   SAMPLE 1.

Fitting these together, we see that the general appearance of the graph must be as follows:

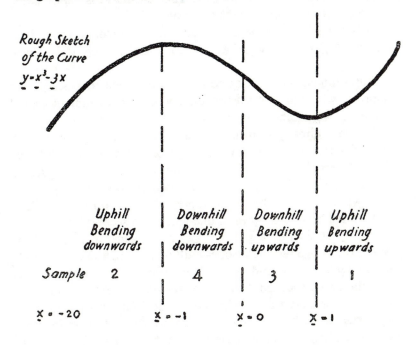

Rough Sketch of the Curve $y = x^3 - 3x$

Uphill Bending downwards	Downhill Bending downwards	Downhill Bending upwards	Uphill Bending upwards
Sample 2	4	3	1
$X = -20$	$X = -1$	$X = 0$	$X = 1$

It would, of course, be possible to make an exact graph of the curve, by working out a table and plotting a large number of points on the curve. We should, of course, get the same curve in the end. But usually the method just explained, using y and y'', is shorter, more instructive and more artistic. Some of the examples at the end of this chapter are intended to illustrate this point.

We saw in Chapter 10 that y' measured the *speed*, y'' the *force acting* on a moving body. When y'' was $+$, it meant that the body was being pushed *upward* (or, in some cases, *forward*). When y'' was $—$, it meant that the body was being pulled *downward* (or, in some cases, *backward*).

But, by examining the graph that represents the motion of the body, we can see how y' and y'' are behaving. In this way, by looking at the graph, we are able to say (for example), 'Here the graph is rising very steeply. The body must be moving very fast. But the curve bends downwards. That means y'' is *minus*, some force is putting a brake on the motion.'

It is quite easy, with a little practice, to tell from a graph how the quantities y' and y'' behave, where y' is $+$, where y'' is minus, etc.

But suppose we are not content with a general description: suppose we want to measure the speed at a certain moment? How can we go about this?

We have already seen (in Chapter 10) that the true speed of a body does not differ much (as a rule) from the average speed over a short period of time. If we know how far a body goes in one tenth of a second, we can get *some* idea of how fast it is going.

If we are shown the graph representing an object's motion, can we tell how far it goes in one-tenth of a second?

In Fig. 10 part of a graph is shown, considerably magnified. The distance AB represents one-tenth of an inch, and corresponds to one-tenth of a second. At the beginning of this length of a second the pencil touched the paper at the point C. At the end of the tenth of a second it touched the paper at D. It must have moved upwards through a distance DE during this interval. In other words, DE represents the change in y: that is, Δy. The time that

ROLLING A CRICKET PITCH

The curve on the right records the motion of the roller. To compare this curve with the arrangement of Fig. 9, it would be necessary to turn the page on its side; the roller would then seem to be moving *upwards* (like the pencil-point in the slit).

The motion may be divided into three sections, A, B, and C.

(a) The man has to push hard to get the roller moving. He is pushing *forwards* ($y'' +$). But the roller is not yet moving fast (y' not large, curve not steep).

(b) The roller is now under way. The man walks beside it, but lets it roll without pushing or pulling ($y''=0$. *No* force).

(c) To stop the roller running into the wickets, the man has to *pull* it *back* ($y''-$).

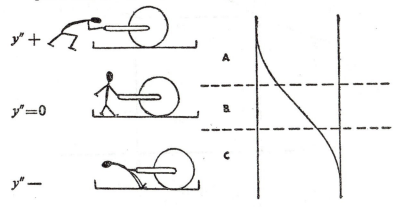

Note that the man is working hardest in sections A and C, but the roller is moving fastest in section B. The *greatest force* (y'' large) does not occur at the same time as the *greatest speed* (y' large).

One can also test this with a bicycle. Use the highest gear, and note that (on a calm day) one works hard in *getting* the bicycle moving, not in *keeping* it moving.

has passed is represented by AB. Thus AB is Δx. The average speed, $\dfrac{\Delta y}{\Delta x}$, is therefore found by dividing the length DE by the length AB. AB is the same length as CE. DE divided by CE therefore represents the average speed. And it is clear that DE divided by CE gives us a rough measure for the steepness of the curve between C and D.

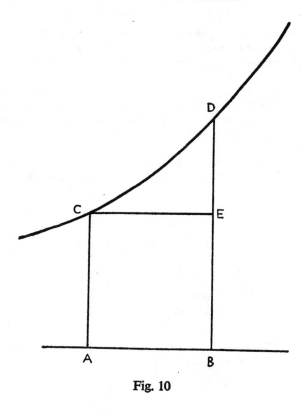

Fig. 10

If instead of a tenth of a second we had taken a hundredth or a thousandth, we should have found a better answer – one nearer to the exact value of y'. So that it is possible to explain what y' is without bringing in the idea of speed at all. Starting with the graph, one takes a point B near to A, draws the figure, and measures DE and CE. One then works out $\dfrac{DE}{CE}$. Then start again: take B still nearer to A, and work out DE divided by CE for the new figure. As B keeps on getting closer and closer to A, the answer keeps on getting closer and closer to some number. The number which it approaches is y': we may regard y' as measuring the steepness of the curve at the point C.

In this way we have given a meaning to y', quite apart from the idea of movement. The same can be done for y'', and these symbols (as was mentioned earlier) can be applied to problems about the shape of a hanging chain, or the arch of a bridge or the best curve for the tooth of a gear-wheel. It need not surprise you if you find still other applications of y' where neither shape nor speed seems to come in. For instance, x might represent the temperature of a body, and y might measure the amount of heat put into the body. It is then possible to give a meaning to y'. Whenever you meet the symbol y' or $\frac{dy}{dx}$, there must be in the problem two quantities, x and y, which are connected with each other, so that any change in x automatically produces a change in y.

So that our definition of y' as a *speed* may be regarded as a kind of scaffolding. It was helpful at the beginning, but no one keeps the scaffolding up when the house has been built. When you have had some experience of the different applications of y' you will probably be able to think of it simply as the number which $\frac{\Delta y}{\Delta x}$ approaches when Δx is made smaller and smaller. But do not be in a hurry to think of it in this way. Even when you are thoroughly familiar with the idea of y', it will often help you to think of it as a speed, or as the steepness of a graph.

The Use of Rough Ideas

It may not have struck you before what a subtle idea 'speed' is. It was quite easy to explain 'average speed'. A car has gone 30 miles in an hour: its average speed has been 30 m.p.h. This is a simple enough idea. But its speed at the precise moment of a collision? That is a much more difficult idea: it is a much harder thing to measure. We have to perform quite a complicated process. Work out the average speed for the minute before the collision, for the second before, for one-tenth of a second before, and so on. See if the answers approach closer and closer to some number; if so, this number represents the speed at the moment of collision.

Actually to carry out this process, we should have to measure very short periods of time – in turn, a hundredth, a thousandth, a millionth of a second – and the very short distances passed over in these times. And even then we should not be quite sure of getting an exact answer. It would always be possible that in the last billionth of a second the driver pressed the brake harder than before, so that the speed at the last instant was really less than we should have expected from the average speed over the last millionth of a second.

Engineers and pure mathematicians differ in their attitude to this question. The engineer thinks the discussion is a waste of time. He does not mind whether the speed was 50 m.p.h. or 50.00031 m.p.h. He cannot measure speed beyond a certain degree of accuracy, and he does not want to, anyway. Even if the brake is put on harder in the last billionth of a second, it will alter the speed by only a tiny amount. So far as an engineer is concerned, the average speed for the last millionth of a second *is* the speed at the moment of collision.

Why does the pure mathematician not agree with this point of view? It is not entirely due to the passion for hair-splitting which affects some mathematicians. One reason is historical. At first, calculus was treated rather in the spirit of the practical man. Very small periods of time were considered. In working out the average speed, this period of time was supposed to be some definite number, bigger than nothing. But in the answer certain parts came which were not wanted, and mathematicians then turned round and said that the little period was so small that it could be treated as nothing. So that at one time a millionth was a millionth, and at another time (when it gave a more convenient answer!) a millionth was treated as being nothing. Naturally students felt that this was a queer subject. Some mathematicians refused to believe that true results could follow from such a method. So mathematicians were forced to clear up the confusion, and to find a more exact and logical way of explaining what they meant by speed. In modern books on the calculus, written for professional mathematicians, you will therefore find very long and careful proofs, written in exact, lawyer-like fashion. It is good to understand these proofs, but not when starting calculus. First learn to use calculus, to see what can be done with it, to feel what it is about. In the course of this you will gradually become aware of a need for more exact ideas – then is the time to study the modern treatment, usually known by the name of *Analysis*.

There are other reasons for using the exact speed. For one thing, practical men are not agreed on the shortest amount that need be considered. The carpenter works in hundredths of inches, the engineer in thousandths, the scientist in millionths – microbes, atoms, rays of light. For a locomotive engineer one hundredth of a second is a short time; for the radio engineer, who thinks in terms of so many million cycles a second, a hundredth of a second is an eternity. The pure mathematician, whose results may be used by any of these men, can be sure of satisfying every possible demand only by giving the *exact* result.

Again, the exact result is often simpler than the inexact one. When we studied the speed of a body corresponding to the formula $y = x^2$, we found several rough results. For instance, we found that the speed after one second was between $1 \cdot 99$ and $2 \cdot 01$. Suppose we had stopped here, and said $2 \cdot 01$ is good enough as an answer. This is a more complicated result than the exact value 2. If we had throughout treated one-hundredth of a second as a sufficiently short time for our purposes, we should have been led to the formula $y' = 2x + 0 \cdot 01$, roughly. This is more complicated than the exact answer, $y' = 2x$. Even engineers use $2x$ as the speed corresponding to x^2. The simpler answer makes up for the more complicated definition.

There is, as you see, much practical justification for the exactness beloved of pure mathematicians. But this is only one side of the question. There are often cases in which the rough idea is very helpful. Often the rough idea of a problem will enable us to see what the problem means, to see the way to a solution. Our answer may be incorrect by a few millionths, but it will be sufficient to give us a general idea of the solution. We may then be able to go over our working, and polish each stage of the work, until the whole thing becomes exact. Or we may rest content with the rough solution of the problem. Many problems which are difficult by exact methods are studied by research workers, by rough methods. The answers are given true to two decimal places, or whatever it may be, sufficient for the purpose required.

Some Examples of Rough Ideas

Suppose, for instance, we were given the problem of finding $\dfrac{dy}{dx}$ (that is, y') corresponding to the formula $y = \log x$. This is a new problem. We know how to deal with an expression built up from powers of x, but $\log x$ does not belong to this simple type. What is to be done?

Students who have learnt merely to do simple problems by text-book methods are, of course, helpless in face of a new kind of problem. They sit before it, and do nothing at all. I hope readers will not find themselves in this situation, but that they will already see how to experiment with this problem, how to seek for a solution.

Do we know what $\log x$ is? If you had difficulty with Chapter 6 it is not the slightest use going on with this chapter. Obviously, if you have no clear picture of the meaning of $\log x$, it is absurd to expect to reason correctly about the speed of growth of $\log x$. If necessary, then, read Chapter 6 again. Get a table of logarithms, and draw a graph to illustrate the formula $y = \log x$. (By $\log x$ I mean the ordinary logarithm as given in the usual tables. $\mathrm{Log}_{10} x$ is the complete sign. This means – to use the language of Chapter 6 – that 'one complete turn' corresponds to multiplication by 10.) Draw this graph for values of x between 1 and 10, plotting sufficient points to give an accurate graph. You will notice that the graph is most steep at $x = 1$. The larger x gets, the less steep the graph. We shall expect for y' some formula which makes y' get less as x gets larger.

We can obtain a *rough* idea of y' by taking a change of $0 \cdot 1$ in x, and seeing what change this produces in y. Part of the work is shown below. Instead of writing the values of x in rows across the page, it is more convenient to set the work out in columns.

TABLE XIII

x	$y = \log x$	Δy	$\dfrac{\Delta y}{\Delta x}$
1	0·0000	0·0414	0·414
1·1	0·0414	0·0378	0·378
1·2	0·0792	0·0347	0·347
1·3	0·1139	0·0322	0·322
1·4	0·1461	0·0300	0·300
1·5	0·1761	0·0280	0·280
1·6	0·2041		
. . . .			
2·0	0·3010	0·0212	0·212
2·1	0·3222		
. . . .			
10·0	1·0000	0·0043	0·043
10·1	1·0043		

In the first column are values of x. Each number exceeds the previous one by 0·1. So the change in x, Δx, is always 0·1. The second column gives the logarithms of the first column, found from a table of logarithms. The third column gives the changes in the second column, Δy. The fourth column gives a rough estimate of the speed, y'; the change in y, Δy, has been divided by the corresponding change in x, Δx. Dividing by 0·1 is the same as multiplying by 10. The numbers in the fourth column are thus ten times those in the third. The table given above is not complete: the numbers between 1·6 and 2·0 and those between 2·1 and 10 have been omitted, to save space. These gaps should be filled in by the reader.

The numbers in the fourth column give us a *rough* measure of the steepness of the graph of $y = \log x$. It is clear that the graph becomes less steep as x increases – a fact we have already noticed. The next problem is to find a formula for these numbers. Since the numbers are not exact, we shall be content with a formula that fits them reasonably well – we do not expect an exact fit.

Guessing a formula is often difficult. A new discovery usually requires years. No one should be discouraged if he takes a few weeks to solve a problem of this type. One simply has to try one idea after another until one hits on the right result. It helps if one makes a collection of graphs of different functions. One can draw the graph of the table given, and see which graph it most resembles.

One may find a clue, in our problem, by comparing the result for $x = 1$ with that for $x = 2$. Opposite $x = 1$, we have in the fourth column $0 \cdot 414$. Opposite x $= 2$ we have $0 \cdot 212$. Now $0 \cdot 212$ is roughly half of $0 \cdot 414$. This suggests that $x = 3$ will correspond to one-third of $0 \cdot 414$, $x = 4$ to one-quarter, and so on. $x = 10$ should correspond to one-tenth of $0 \cdot 414$ – that is, $0 \cdot 0414$. The table gives $0 \cdot 043$, which is not much different. To $x = 1 \cdot 5$ should correspond $0 \cdot 414$ divided by $1 \cdot 5$ – that is, $0 \cdot 276$. The table gives $0 \cdot 280$, which is quite close – as close, at any rate, as we can expect with such a rough method.

This work therefore suggests that y' corresponding to $y = \log x$ is *something like* $\dfrac{0 \cdot 414}{x}$.

This result should be regarded as a sort of hint. It suggests that we go back to the explanation of logarithms given in Chapter 6, and try to see if there is some obvious reason why log x should grow at a speed which is proportional to $\dfrac{1}{x}$. Actually, there is, and it can be shown that $\dfrac{0 \cdot 434294}{x} \cdots$ is the true answer to the problem. Our rough method has shown us the *form* of the answer, and has brought us reasonably near to the true value.

(In Chapter 6 we explained the meaning of 10^x. What is the formula for y', if $y = 10^x$?)

You may remember that it was mentioned in Chapter 6 that a slide-rule could be made to any scale we liked. If we mark the number 10 one inch from the end of the scale, the number x occurs at the distance log x inches from the end. But we could take 10 at any other distance, and we could still make a perfectly good slide-rule. Our result for y' corresponding to $y = \log x$ suggests that it might be worth while to alter the scale in a particular way. We found that y' was equal to $\dfrac{0 \cdot 434294 \cdots}{x}$. If we took, instead of $y = \log x$, the formula $y = \dfrac{\log x}{0 \cdot 434294 \cdots}$ we should have a simpler result: y' would then be simply $\dfrac{1}{x}$. This new expression y will do just as well for the distance at which the number x has to be marked. If we mark every number x at a distance $\dfrac{\log x}{0 \cdot 434294 \ldots}$ inches from the end of the scale, we obtain a slide-rule that is on a larger scale, but is otherwise no different from the previous one. 10 now occurs at a distance equal to $\dfrac{\log 10}{0 \cdot 434294 \ldots}$ As log 10 is 1, this can be worked out; it is equal to $2 \cdot 30258 \ldots$ inches. The number which now occurs at a distance of one inch is $2 \cdot 71828 \ldots$ This number is important in mathematics: it is given a special name, and is always spoken of as e. The distance at which any number occurs on this new slide-rule is called the *Natural Logarithm* of the number. The natural logarithm of x is written $\log x$.

When we first explained logarithms in terms of ropes wound on posts, we took the effect of one complete turn to be 10. The only reason for doing this was the accidental fact that we have ten fingers. If we had eight fingers, we would probably take one turn to correspond to 8, and we should get just as good a table of logarithms. For these we should use the sign $\log_8 x$. Any other number could be used. It need not be a whole number. $2\frac{5}{8}$ would do, for instance. All these different numbers would lead to perfectly good slide-rules, but of different sizes. We should always find $y' = \dfrac{a}{x}$, where a stands for 'some fixed number'. It is natural to prefer that system of logarithms for which $a = 1$. For this reason, in theoretical work we usually employ $\log_e x$, the 'natural logarithm'. If $y = \log_e x$, $y' = \dfrac{1}{x}$.

The Cartwheel Problem

We now consider another problem, in which a rough idea is helpful. If a wheel – for instance, a cartwheel – is rolling along a

flat road, how fast do the various parts of it move? They certainly
do not move all at the same speed. You will often notice, when
a motor-car passes, that the lower spokes can be clearly seen, but
the top spokes move so fast that they are invisible. How is this
to be explained?

Many people find a rolling wheel difficult to imagine – not, of
course, difficult to imagine in a vague way, but difficult to imagine

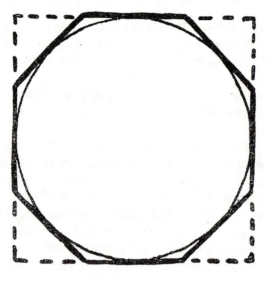

Fig. 11

so clearly that the speed of each part can be seen. Let us therefore
replace this problem by a simpler one. It is fairly easy to imagine
a square rolling, as, for instance, a large log of square section
being pushed along the pavement. It starts with one side flat on
the pavement. Then it turns about one corner, until the next side
is flat on the pavement. Then it turns about the next corner, and
so on. It is easy to see that the corner at the bottom of the square –
the corner about which the whole thing turns – is at rest. The
farther away a point is from this corner, the quicker it moves.

Now let us make our square log rather more like a circle, by four straight cuts with a saw, to remove the four corners of the square, as in Fig. 11. It is still quite easy to imagine the motion. As before, the point which touches the pavement, when the figure rolls, is (at any moment) at rest.

We can continue in this way, cutting off corners, and making the figure more and more like a circle. It will never *be* a circle, but it can be made *as near to* a circle as we like. The figure with 128 corners would make quite a good wheel for most practical purposes.

We are thus led to the conclusion – which every engineer has to know – that a rolling wheel turns about its lowest point, which is (for an instant) at rest.

The same method of approach would enable us to see what curve any point of the rolling wheel describes. The curves known as cycloids, epicycloids and hypocycloids arise in this way.

In the cutting of gear-wheels, a special curve has great advantages. This is the curve described by the end of a cotton thread, as the thread unwinds from a fixed reel. It may help you to see just what the thread does, if you think of the reel as being, not an exact circle, but a figure with corners such as we had above. The more corners you imagine, the nearer you come to the true state of affairs.

In the same way, the motion of a small body rolling down a curved hill can be simplified: we can replace the curve by a figure with corners. If we know all about the behaviour of a small body rolling down a straight line, we can thus build up a picture of a body rolling down a curve.

Again, we can study the behaviour of a hanging string by considering what happens to a hanging chain, made up of straight links like a bicycle chain. The more links we imagine, the nearer we come to a true idea of the string.

All these illustrations use what mathematicians call the idea of a Limit. Mathematicians speak of a circle as being the *limit* of the figures with corners described above: this simply means that you can get *as near as you like* to a circle by making *a sufficiently large number* of straight saw cuts. You can get *as near as you like* to the curve of a string by taking *sufficiently short* links for your chain.

You can get *as near as you like* to the speed $\frac{dy}{dx}$, by taking *sufficiently small* Δx and calculating $\frac{\Delta y}{\Delta x}$.

ILLUSTRATIONS AND EXPERIMENTS

1. The movements of trains on railways are recorded on a graph, when time-tables are being worked out, or emergency trains are fitted in (See J. W. Williamson, *Railways Today*, for a reproduction of such a graph.)

Draw graphs showing the position of the following trains:

    (i) Leaves London at 12.00; travels at steady speed of 20 m.p.h.

    (ii) Leaves London at 12.00; travels at 30 m.p.h., but stops for 5 minutes every half-hour.

    (iii) Leaves London at 12.48 and travels at 50 m.p.h.

In these graphs, have distance going *upwards*, the passage of time towards the *right*. See for yourself that the faster trains have steeper graphs, and that this makes it possible for the express (Number iii) to overtake the goods train (Number i).

2. A cricket ball is thrown straight up into the air. After x seconds its height is y feet, where $y = 40x - 16x^2$. Work out y' and y''. Show that y'' is always minus, and that y' starts by being $+$, but later becomes $-$. At what time is $y' = 0$? Draw the graph for the motion of the ball. When is the ball at its greatest height? What is this height?

3. A piece of wood is hurled downwards into the sea. After 3 seconds it again appears on the surface. A scientist states that during this period its graph had the equation $y = -30x + 10x^2$, where x stands for the time in seconds after it entered the water, and y for its height in feet (measured upwards, so that depths below sea-level are minus). How far is this a reasonable description of what the piece of wood does?

4. A body is shot from a catapult (it does not matter whether it is a pebble or a glider), and its motion is recorded by means of

a graph. How would you expect this graph to bend (i) while the body is still in the catapult and gathering speed; (ii) after it has left the catapult?

5. Find out the general appearance of the graphs corresponding to the formulae below, by working out y' and y'' and seeing what signs ($+$ or $-$) they have. Consider x from -10 to 10.

(i) $y = x^2 + x$. (ii) $y = x^3$. (iii) $y = x^3 - x$. (iv) $y = x - x^2$. (v) $y = x^3 - x^4$.

Be careful to note that where $y' = 0$ the curve is, for an instant, *level*. This applies particularly to (ii) and (v).

6. The following example can be dealt with by the method of Rough Ideas.

A man is walking along a straight path, which points due East, at a speed of 5 feet a second. The man whistles to his dog, which is at that moment 30 feet due North of the man. The man does not stop walking. The dog runs, at a speed of 20 feet a second, always facing towards his master. Draw roughly the curve which the dog describes.

Position of the dog at intervals of $\frac{1}{10}$ second.

Position of man at intervals of $\frac{1}{10}$ second.

About how long does the dog take to reach the man? (It is clear that if the dog had given a little more thought to what it was doing it would have run in a straight line, towards a point somewhat ahead of its master, instead of behaving in this way.)

Method. Replace the steady movement of the man by a series of jerks. Suppose the man to remain still for one-tenth of a second, then suddenly to shoot forward six inches. In this way the man will cover 5 feet a second. The dog will run in a chain of straight lines, each 2 feet long, pointing in turn towards the various positions in which the man stands. Drawing this chain we can see roughly how the dog moves and how long he takes to reach the man.

CHAPTER 12

OTHER PROBLEMS OF CALCULUS*

'One of the hardest tasks that an expert in any subject can undertake is to try to explain to the layman what his subject is, and why he makes such a fuss about it.'
– G. C. Darwin, Introduction to *The Story of Mathematics*, by D. Larrett.

THE pioneers of every subject are amateurs. They start with the same knowledge and the same methods of thinking as any other uninstructed person. The early discoveries in any branch of science can be explained in everyday language, and often seem so obvious that the science appears hardly worth studying.

Later generations, building on the work of the pioneers, study more and more complicated problems, and in the course of this introduce new ideas, new words, technical terms which are Greek to the layman. The later discoveries of any science have to be stated in technical terms: they appear remote from ordinary life, and extremely hard to understand. The science now seems so difficult that it appears not worth while to try to master it.

In mathematics, as in other sciences, each generation builds upon the foundation provided by past workers, and adds another storey. By now the building is something of a skyscraper. There are many remarkable books on the eighteenth floor, but they are written in a language which is comprehensible only to those who are thoroughly familiar with the works on the seventeenth floor – and so on, floor by floor, until one reaches the ground floor and the multiplication table.

No living person is familiar with all the mathematical discoveries that are stored in the libraries of the various learned societies throughout the world. Every mathematician has to find out for himself just which parts of the subject are useful for his own purposes, which technical terms and ideas he has time to become familiar with, and to apply to his own problems.

* This chapter may be skipped by those who find it hard.

In this chapter a number of processes will be discussed which are useful to those engaged in the more exact sciences – physics, chemistry, engineering – and to those who use any kind of machine, from a drill to an aeroplane. This type of mathematics is also creeping in to subjects such as biology, economics, even psychology. If you are not interested in such applications, you will perhaps not find this chapter of interest, nor will you find much point in learning calculus at all, unless you belong to the type which is interested in mathematics for itself. If you neither like, nor need, calculus it is a sheer waste of time to study it.

In the course of scientific work it is frequently necessary to find $\dfrac{dy}{dx}$ for expressions $y\,(x)$ which are more complicated than any we have so far considered. One might, for instance, come across an expression such as $y = \left(\dfrac{x}{x^2 + 1}\right)^3$, and wish to know y' corresponding to this. Here a whole series of processes has been carried out. If we start with x, we have to work out x^2 and add 1 to this, giving $x^2 + 1$. Then x has to be divided by this result. The answer to this has to be raised to the power 3.

The problem is dealt with by splitting it up. We bring in new letters, and break up the chain of processes. In calculating y we were first led to calculate $x^2 + 1$. We will call this first result u. $u = x^2 + 1$. How quickly does u grow? We know this from our earlier work: $u' = 2x$. We then calculate $\dfrac{x}{x^2 + 1}$ ($x^2 + 1$ being the same thing as u). Call this result v, so that $v = \dfrac{x}{u}$. Now we know that u grows at the rate u', and that x grows at the rate 1. v is obtained by dividing x by u. Since we know x and u, and we know how fast each of them is growing, it ought not to be too difficult to find how fast v is growing. Suppose we solve this problem, and find the answer, v'. We now come to the final stage. y is obtained by raising v to the power 3; that is, $y = v^3$. y is v^3 and v is growing at the rate v'. How fast is y growing?

So far the problem has not been solved. We have simply shown that the one complicated problem can be split up into three simpler ones: I. To find u' when $u = x^2 + 1$. II. To find v' when $v = \dfrac{x}{u}$ and u' is known. III. To find y' when $y = v^3$ and v' is known.

It is because complicated problems can be split up in this way into simple ones that you find certain theorems in every calculus text-book, dealing with the differentiation of a Sum, Product, Quotient, and the Function of a Function. All these theorems have an object – to enable you to find y' corresponding to any formula, however complicated, by splitting the problem into simpler ones.

Differentiation of a Sum

Consider an example – rising prices. Let £y be the price of a watch after x days of war (rising at the rate y') and £z the price of a chain (rising at the rate z'). How quickly does the price of a watch and chain rise? Clearly $y' + z'$. As the price is £$(y + z)$, this shows how easy it is to find the rate of increase of the sum of two changing quantities.

Differentiation of a Product

Let n be the number of men in a town, and p the number of pints drunk daily by each man. Then np is the total number of pints drunk. If n is increasing at the rate n' and p at the rate p', how fast is np increasing? The answer is $p'n + n'p$.

Differentiation of a Quotient

If b barrels of beer are provided for n men, each will receive $\dfrac{b}{n}$ of a barrel. If the number of men increases at the rate n' and the number of barrels at the rate b', how fast does $\dfrac{b}{n}$ change? The answer turns out to be $\dfrac{b'n - n'b}{n^2}$. Notice how this answer compares with common sense. If $n' = 0$, it means that the number of men stays the same, and if b' is $+$, it means that the number of barrels is increasing. In that case, $\dfrac{b}{n}$ is increasing, and its rate of change should be $+$. This is so, for the formula above. If, on the other hand, the number of barrels stays the same, $b' = 0$, while the number of men increases, so that n' is $+$, the formula becomes $-\dfrac{n'b}{n^2}$, which has a minus sign, as it ought to do: the share per man is getting *less*, the change is for the worse, a *minus* sign is to be expected.

Function of a Function

It is well, at this stage, to look back to page 96 and to read again the sentences explaining what is meant by y being a function of x; namely, that y is connected with x by some rule. What now is 'a function of a function'? Consider the formula $y = \log_e (x^2 + x)$. We could make a table of y in the following way. In the first column we could enter the numbers, x. In the second column we could enter the corresponding numbers, $x^2 + x$. In the third column we could put the logarithms (to base e) of the numbers in the second column. This third column would then give the numbers $\log_e(x^2 + x)$. We have x in the first column, y in the third column. Let us call the numbers in the middle column z. The numbers in the second column are found from those in the first column by a definite rule. So z is a function of x. The numbers in the third column are found by a definite rule from those in the second. So y is a function of z. It is this process which gives rise to the name 'function of a function'. In fact, $z = x^2 + x$ and $y = \log_e z$.

Now we know all about the rule connecting x with z, and we know all about the rule connecting z with y. It ought not to be too difficult to find how fast y increases.

It is possible to illustrate this double connexion by a machine. The relation $z = x^2 + x$ can be expressed by means of a graph. In Fig. 12 the curve OB represents a groove cut to the shape of this graph. OA and OC are straight grooves. A represents a small piece of metal sliding in the groove OA. In the same way B

slides in the groove OB, and C in the groove OC. A small ring is fixed to B, and through this ring pass the two rods, AB and CB, which are soldered to the sliding pieces A and C in such a way that AB is always upright, and BC is always level. If A moves, B is forced to move, and this in turn forces C to move. The distance OA represents x, the distance OC represents z. Any change in x produces a change in z, and the machine is so designed that $z = x^2 + x$.

In the same way we can express the connexion, $y = \log_e z$. y is represented by the length OE. z is already shown by OC. The curved groove GD is the graph of $y = \log_e z$. The rod CD is always level, while ED is always upright. Both pass through a ring at D.

The whole chain of events can now be seen. If x (OA) changes, z (OC) cannot help changing, and because OC changes, OE (y) must also change.

How fast? We know that z increases $\dfrac{dz}{dx}$ times as fast as x. y increases $\dfrac{dy}{dz}$ times as fast as z. So y must increase $\dfrac{dy}{dz} \cdot \dfrac{dz}{dx}$ times as fast as x. That is, $\dfrac{dy}{dx} = \dfrac{dy}{dz} \cdot \dfrac{dz}{dx}$. This is the theorem about a Function of a Function and its rate of change.

In our particular example it is easy to find $\dfrac{dy}{dx}$. We have to find $\dfrac{dy}{dz}$ and $\dfrac{dz}{dx}$ and multiply these together. There is no difficulty with $\dfrac{dz}{dx}$. $z = x^2 + x$, so $\dfrac{dz}{dx} = 2x + 1$. Now for $\dfrac{dy}{dz}$. $y = \log_e z$. We saw in Chapter 11 that $\log_e x$ grew at the rate $\dfrac{1}{x}$. 'The natural logarithm of any number grows at a rate given by dividing one by that number' is this formula in words. It makes no difference if we call the number z instead of x. So $\dfrac{dy}{dz} = \dfrac{1}{z}$. Accordingly

$\frac{dy}{dx} = \frac{1}{z}(2x + 1)$. But z is short for the number in the second column, $x^2 + x$. This answer is therefore the same as $\frac{2x + 1}{x^2 + x}$ and this formula is the solution of the problem.

By combining the results stated above, it is possible to find $\frac{dy}{dx}$ for very complicated formulae.

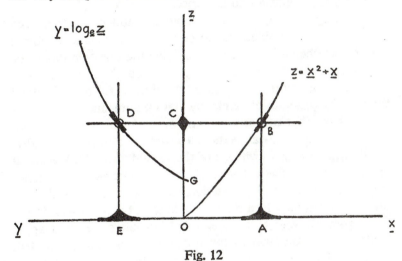

Fig. 12

Integration

We have already considered the problem of differentiation – that is, the problem of finding the speed y' of a body which moves in a manner described by means of a formula for y. The opposite problem frequently occurs: we know the speed at every moment, we have to find how far the body goes after any number of seconds. In other words, we are given a formula for y', and we are asked to find a formula for y. This is the problem of *integration*.

y' need not, of course, be considered as the speed of a moving body. It represents the rate of change of y, whatever y may be. For instance, it is easy to discover *how quickly* the pressure increases as a diver goes deeper into the sea. So y might represent the pressure per square foot on a diver's helmet, when he is at a depth x feet. It is easy to find the rate of increase, y'. In order to find y, the problem of integration (very easy in this particular case) has to be solved. Integration is also used in connexion with the question of air pressure at different heights, a question of interest to mountaineers, airmen, weather experts, and others. There are few, if any, branches of science and engineering in which problems of integration do not arise.

In dealing with any practical problem, a student has to do two things. First, the problem has to be put into mathematical form: then the mathematics necessary to solve the question has to be carried out. The second part is no use without the first. Our study of integration will therefore have two objects: (a) to understand the nature of integration so clearly that we immediately recognize any problem that can be solved by means of integration, (b) to master the mathematical method. The first part, (a), can be understood without any knowledge of the second part, (b). We are at present mainly concerned with (a), though some passing reference may be made to (b). We shall consider a very simple problem – one which can be solved mathematically in two lines – and look at it from all angles. To this simple problem we shall apply methods capable of solving far more difficult questions: we shall use steam hammers to crack a walnut, in fact. The object of this will not be to crack the walnut, but to demonstrate how the steam-hammers work. Our simple problem is the following: if $y' = x$, find a formula for y. This problem is not quite complete, as it stands. We are given y' which might represent the speed of a body after x seconds. Obviously, we must know where the body starts, if we are to work out its position. We will suppose, then, that we are also told that $y = 0$ when $x = 0$. The problem is quite definite. We can think of y as the distance of the body from a fixed point P. To begin with, the body is at P, since the distance y begins by being zero. Then the body begins to move. After 1 second its speed is 1 foot per second; after 2 seconds, its speed is 2 feet per second, and so on. The speed does not grow by jumps, but steadily. For $y' = x$ tells us that the speed after $1\frac{1}{2}$ seconds is

1⅓ feet per second; after 1¼ seconds, y' is 1¼, and so on. We have a complete picture of the motion.

The Method of Rough Ideas

Let us try, first of all, to get a rough idea of the distance which the body would go, for example, in the first second. We will split the second into ten equal parts, and see how much we can find out about the distance which the body travels in each tenth of a second. In the first tenth of a second the body moves with a speed which increases steadily from 0 at the beginning to 0·1 at the end. So the average speed lies between 0 and 0·1. So the distance gone must be more than 0 times 0·1, but less than 0·1 times 0·1. We can apply the same argument to each of the other parts. The distance gone in 0·1 second is more than 0·1 times the least speed, and less than 0·1 times the greatest speed in that part of the motion. We can put the argument in the form of a table.

TABLE XIV

Time	Least Speed.	Highest Speed.	Distance gone At least.	At most.	Difference.
0 to 0·1	0	0·1	0	0·01	0·01
0·1 to 0·2	0·1	0·2	0·01	0·02	0·01
0·2 to 0·3	0·2	0·3	0·02	0·03	0·01
0·3 to 0·4	0·3	0·4	0·03	0·04	0·01
0·4 to 0·5	0·4	0·5	0·04	0·05	0·01
0·5 to 0·6	0·5	0·6	0·05	0·06	0·01
0·6 to 0·7	0·6	0·7	0·06	0·07	0·01
0·7 to 0·8	0·7	0·8	0·07	0·08	0·01
0·8 to 0·9	0·8	0·9	0·08	0·09	0·01
0·9 to 1·0	0·9	1·0	0·09	0·1	0·01
		Total	0·45	0·55	0·10

In the first column we have the ten parts into which the first second is divided. Then follow the least speed and the greatest speed in each part of the motion. We then have two columns showing 0·1 times the least speed, and 0·1 times the greatest speed. In each tenth of a second the body must have gone more than the former, less than the latter. The last column shows the difference between the two previous ones. For instance, we know that, in the time between 0·6 and 0·7, the body goes at least 0·06, at most 0·07. The difference between these is 0·01, so we are left uncertain about the distance gone in this time, to the extent of one-hundredth. The total distance gone in the first second is found by adding up. It must be more than 0·45 foot, less than 0·55 foot.

We now have a rough idea how far the body goes in the first second. The difference between 0·45 and 0·55 is 0·1. This uncertainty of 0·1 is due to the uncertainty of 0·01 in each of the ten rows. If we want a more accurate answer, it will be necessary to carry through the same process, using smaller intervals. We might, for instance, divide one second into 100 parts and then carry out a similar calculation, though of course this would be rather long and boring to do. How close would such a method bring us to the true answer! The difference between the highest and the least speed in any small part would be 0·01, instead of 0·1. The distance gone in 0·01 second is found by multiplying the speed by 0·01. So the uncertainty in the distance gone, in any hundredth of a second, would be 0·01 times 0·01 – that is, 0·0001. But there would be a hundred rows in the table (instead of ten) and the uncertainty in the total distance would be 100 times 0·0001, that is 0·01. (Actually we should find that the distance was more than 0·495 and less than 0·505.) This result is ten times as good as the time before – we are repaid for having to do ten times as much work to obtain it. By taking still more intervals, we could get still better results.

This method is used only when a problem is so difficult that no other method will work. Even then, some shortening of the work would be arranged. The method is not given here as a good way of finding the actual answer, but rather to show what the problem means. The process above will help you to understand the sign used for integration. We have already used the sign $\triangle x$ for the change in x. In the table given above, each row of the first column represents a change of 0·1 – e.g., 0·7 to 0·8. The next two columns tell us the least speed and the highest speed – that is, they help us to see how large y', the speed, is, during this part of the time. In the same way, the fourth and fifth columns give us a number rather less, and a number rather more, than the distance gone during any small part of the time. As distance gone is measured by the speed (y') times the time that passes (Δx), we may think of these columns as representing $y' \, \Delta x$. Of course there is some doubt about the meaning of y': for instance, as x goes from $0 \cdot 6$ to $0 \cdot 7$, y' also goes from $0 \cdot 6$ to $0 \cdot 7$, and it is not clear whether we should take y' as being $0 \cdot 6$ or $0 \cdot 7$ or some number in between. It is because of this uncertainty that we have the two columns, one headed 'At least', the other 'At most'. This must be borne in mind.

We then estimate the distance gone in the whole of the first second by adding up the fourth and fifth columns. So that the 'sum of $y' \Delta x$' is (at least) $0 \cdot 45$ and (at most) $0 \cdot 55$.

Here we have two estimates – one rather too small, one rather too large. But, fortunately, by taking smaller lengths of time – that is, taking Δx to be $0 \cdot 01$ or $0 \cdot 001$, etc. – these two estimates approach closer and closer to each other. In other words, if Δx is made very small, it matters *very little* whether we take y' to be the highest or the lowest speed that occurs in the interval of time, Δx. The answer will be the same, whichever is taken. If this were not so, we should have to bring in a new sign, such as $y' \Delta x(L)$, to mean 'the *least* speed, y', times the change in x, Δx'. But, as it has turned out, this would be a waste of time. The least speed and the highest speed give answers that draw closer and closer together as Δx gets smaller.

In Chapter 10 it was mentioned that $\dfrac{\Delta y}{\Delta x}$ came nearer and nearer to a certain number as Δx became smaller. The number thus approached was therefore christened $\dfrac{dy}{dx}$. In the same way, the number we have just been estimating – the number which is more than $0 \cdot 45$ and less than $0 \cdot 55$, more than $0 \cdot 495$ and less than $0 \cdot 505$, etc. – is represented by $\displaystyle\int_0^1 y' dx$. The sign $\displaystyle\int$ is an old-fashioned S, S for 'Sum'. The sign is meant to indicate that the number can be found by multiplying y' by Δx for each short part of the time, finding the sum of these, and seeing what happens when Δx gets very small. The numbers 0 and 1 are put in to show that we are interested in the distance gone during the first second – that is, between $x = 0$ and $x = 1$. In other words, the change of x from 0 to 1 has to be split up into little changes Δx, as in the first column of the table. $\int_{2}^{5} y'dx$ would represent the distances gone in the period of time from 2 seconds after the start to 5 seconds after the start. $\int_{0}^{n} y'dx$ represents the distance gone in the first n seconds. Since we have supposed y' to be given by the formula $y' = x$, we may replace y' by x, and write $\int_{0}^{1} x\, dx$. It seems likely, from the work above, that the number we are looking for is $0 \cdot 5$, and it can be proved that this is the correct answer. In symbols, $0 \cdot 5 = \int_{0}^{1} x\, dx$. $\int$ is known as 'the integral sign'. 'To integrate' means 'to make whole': the name, I suppose, is chosen because the process consists in putting together a lot of little bits, all the little changes in y that occur in the brief passing moments of time.

Ways of Seeing Integration

It is useful to know different ways of getting the same result. Then, in any problem, one can imagine the illustration which is most convenient.

We may translate the symbol $\int_{0}^{1} x\, dx$ as the distance gone, in the first second, by a body whose speed is always equal to the number of seconds that have passed (including fractions of a second).

To record such a motion we might use the device illustrated at the beginning of Chapter 11. It would be rather difficult to arrange for the speed y' always to be exactly equal to the number of seconds, x. Let us once more be content with a rough idea. During the first tenth of a second let the pencil-point stay at rest

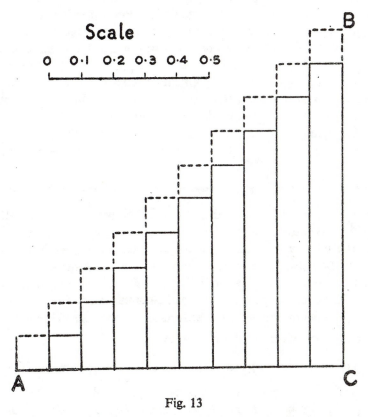

Fig. 13

in the slit *AB*. During the next tenth of a second let it move upwards with a speed of 0·1 foot a second: from $x = 0·2$ to $x = 0·3$,
let the speed be 0·2 foot a second, and so on. In fact, let the speed
during any interval given in the first column of Table XIV be the
number given in the second column of that table. The graph that
results will consist of straight lines, joined together like a chain.
As we saw in Chapter 11, y' measures the steepness of these lines.
As y' increases steadily, each line will be steeper than the previous
one. So we could equally well have explained integration from the
problem: You are told how steep a curve is at every point, draw
the curve.

It is also possible to illustrate Table XIV directly. The numbers
in the fourth column are obtained by multiplying those in the

second column by 0·1. But multiplication – for instance, 0·9 times 0·1 – can be represented by the area of a rectangle, with sides 0·9 and 0·1. The ten numbers in the fourth column can be represented by the area of ten rectangles, as in Fig. 13. The sum of these numbers, 0·45, represents the total area, below the heavy line. In the same figure the area below the dotted line represents 0·55, the sum of the numbers in column five.

Between the dotted line and the heavy line are ten squares, each with an area equal to 0·01. These squares represent the numbers in the last column.

We know that $\int_0^1 x\, dx$ represents a number larger than the area below the heavy line, less than the area below the dotted line. By taking 100 instead of ten steps, we should get an even better idea of the number we want. But however many steps we have, the heavy line always lies below the straight line AB, the dotted line always lies above it. If we draw the line AB, the area of the triangle ABC is always more than the area below the heavy line, less than the area below the dotted line. In fact, the area ABC is equal to the number we are looking for, $\int_0^1 x\, dx$.

This result is quite general. If $f(x)$ is any function of x, then $\int_a^b f(x)\, dx$ will always represent the area below the graph of $f(x)$, between $x = a$ and $x = b$. *The problem of finding the area inside any curve is a problem in integration.* You should try for yourself to draw an area which represents $\int_0^1 x^2\, dx$.

The connexion between integrals and areas is useful in two ways. We can use an area to illustrate the meaning of an integral, and to help us to understand the behaviour of integrals. Secondly, we can find the actual size of a particular area by working out the value of an integral.

A Shorter Method

The integral, $\int_0^1 x\, dx$, can be found with very little work. We started the problem by trying to find y such that $y' = x$, and $y = 0$

when $x = 0$ (see page 166). But we know that to the formula $y = x^2$ there corresponds the speed $y' = 2x$. $2x$ is just twice the answer we want for y'. We can put this right by taking y half as large – that is, we consider the formula $y = \frac{1}{2}x^2$. This gives exactly the right answer, $y' = x$. Also, $\frac{1}{2}x^2$ equals 0 when $x = 0$, so the condition $y = 0$ when $x = 0$ has been met. So $y = \frac{1}{2}x^2$ is the formula we want. It gives the distance y corresponding to x seconds. Putting $x = 1$, we find $y = \frac{1}{2}$. So the distance gone in one second is $\frac{1}{2}$. This agrees with the result $0 \cdot 5$ which we found by the other method.

Many problems of integration can be solved by this method. The idea is simple enough. We have already learnt how to find y' corresponding to many different types of function, y. We are now asked to do the opposite problem: y' is given, we have to find y. It is natural to turn back to our records of the first problem. If in these we find y' of the type required, the problem is immediately solved. For instance: we have shown that to $y = \log_e x$ corresponds $y' = \dfrac{1}{x}$. If we are asked to find $\int \dfrac{1}{x}\, dx$, this is the same as saying: if $y' = \dfrac{1}{x}$, what is y? Obviously, $y = \log_e x$ gives *an* answer to this question. The complete answer will depend on the other condition: it is not enough to know how fast a body moves, one must also know where it is at some instant.

Differential Equations

Very many practical problems lead to what is known as a differential equation. The nature of a differential equation may best be seen by considering a definite example.

The light from an electric lamp spreads out equally in all directions. Often – as, for instance, in making a motor headlamp or a searchlight – this is inconvenient: we would prefer to have all the light coming out in one direction, and this is achieved by placing a reflector behind the lamp. If the reflected light is to come out in a perfect beam, what shape should the reflector be?

It is known how light behaves when it strikes a mirror. If we take a capital V and underline it, thus V̲, we have a rough picture of what occurs. The line represents the mirror: the left-hand arm of the V̲ may represent the light striking the mirror, and the right-hand arm the light bouncing off the mirror. The two arms of the V must make the same angle with the line of the mirror. A billiard ball bounces off a cushion in more or less the same way, if it is free from spin.

We shall not get a proper beam if we simply put an ordinary straight mirror behind the lamp. The reflected light will scatter in different directions, as can be seen by drawing a figure.

We might tackle the problem by taking a large number of short pieces of mirror and trying to join them in a chain, in such a way as to get a proper beam. In Fig. 14 P represents the point where the electric lamp is placed. O is some other point, and we want to obtain a beam of light pointing in the direction OP. OA represents a short piece of mirror, so placed that the light from P which

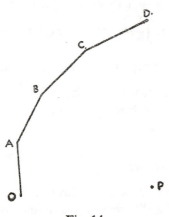

Fig. 14

strikes the mirror at O is reflected back along the line OP. Of course the light from P which strikes the mirror between O and A will be reflected slightly upwards, slightly away from OP, but if the length OA is short, this will not be serious. When we reach A, we join on our next piece of mirror AB, in such a way that the ray of light PA is reflected in the proper direction – that is, parallel to OP. In the same way, we must turn the next piece of mirror, BC, in such a way that the ray PB is reflected parallel to OP. And so the process is continued: each mirror is added in such a way that its lowest point touches the highest point of the mirror before and it is then turned so as to reflect light in the proper direction.

In this way we could build up a mirror which very nearly produces a perfect beam. The shorter the pieces of mirror used,

the better the beam would be. We can easily believe that there is
a curve which reflects the light exactly in the correct direction.
This curve is known as a *parabola*. The mirror built in this shape
is called a *parabolic mirror*. Parabolic mirrors are used in some
types of telescope, and in beam wireless upright wires arranged
in the shape of a parabola are used.

Notice how we built up our chain of lines $OABCD$ ... We
started from O, and then, at each stage, we were told *in what
direction to go*. Any problem which starts from some rule about
the direction to follow at any moment will lead to a differential
equation.

For instance, a ship at sea might steer straight towards a lighthouse. It could start wherever it liked, but once it had started
the direction it had to follow would be fixed. The lighthouse might
be spoken of as a magnet attracting the ship: in the language of
magnetism the ship follows a 'line of force'. The problem would
become more complicated if we had two magnets, each attracting
a moving body. The path of the body would not then be obvious,
as it was for the ship and the lighthouse. Differential equations
therefore appear in the theory of electricity and magnetism.

What does a differential equation look like in algebraic symbols? We have some rule that gives us the direction of the curve
at any point. We might equally well say we have a rule that gives
the steepness of the curve at any point. Now, the steepness of the
curve is measured by y' and the position of a point on a graph is
measured by the two numbers x and y. To every point there corresponds a direction: we might imagine this by supposing the
graph-paper to be covered with little arrows, signposts conveying
the message, 'If you should arrive at this point, depart in this
direction'. By continually following the signposts, one would
follow out some curve. The signposts are arranged *according to
some rule*: if we have any point (corresponding to any two numbers x and y) we have a rule giving the direction of the signpost,
and the steepness of the arrow is measured by y'. So y' is given by
some rule – that is, we have a formula giving y' when x and y are
known.

For instance, if a lighthouse is placed at the point (0, 0), and all ships sail straight towards it, the formula is $y' = \dfrac{y}{x}$. For the steepness of the line joining any point (x, y) to the point $(0, 0)$ is $\dfrac{y}{x}$, and y' must be equal to this.

You will not be able to follow this argument if you are not familiar with Co-ordinate Geometry: you must master the earlier part of Co-ordinate Geometry (the plotting of points, the steepness of straight lines, the angles between straight lines, the distance between two points, the equation of a circle) before you try to learn the theory of differential equations.

EXAMPLES

The treatment of the subjects in this chapter is too sketchy to justify the setting of examples. Readers who have been able to follow the general ideas of this chapter will find examples in any text-book on Differential and Integral Calculus.

CHAPTER 13

TRIGONOMETRY,
OR HOW TO MAKE TUNNELS AND MAPS

'*The utmost care must be taken to avoid errors, and that it is taken is proved by the wonderful accuracy with which the headings driven from opposite ends usually meet ... The Musconetcony Tunnel is about 5,000 feet long. When the headings met the error in alignment was found to be only half an inch, and the error in level only about one-sixth of an inch. In the Hoosac Tunnel, 25,000 feet long, the errors were even smaller.*'
A. Williams, *Victories of the Engineer.*

IN this book I have tried to show (i) that mathematical problems can be stated in the language of everyday life; (ii) that any normal person can think about these problems for himself, by using common sense; and (iii) that the methods given in text-books simply represent the improvements on the original common-sense attack, which have gradually been built up by generations of mathematicians.

In no part of mathematics is this easier to show than in trigonometry. Trigonometry arises from very simple practical problems, such as the building of a railway tunnel, for instance. It may be necessary to make a tunnel which is to come out several miles away, on the other side of a range of mountains, at a point which cannot even be seen from this side. It may be necessary to bore the tunnel from both ends, and to meet somewhere far inside the mountain. How are we to find the correct direction in which to bore?

One method is explained in Chapter 4 of *The Railway*, by E. B. Schieldrop. It is illustrated in Fig. 15. The shaded part represents high ground. It is desired to connect the points *A* and *D* by a

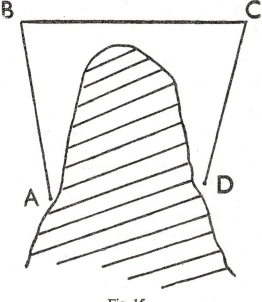

Fig. 15

tunnel. We may be able to find objects B and C, such that B can be seen from A, C from B, and D from C. We measure carefully the length and direction of the lines AB, BC, and CD.

This information is sufficient to fix the position of D. It would enable us to make a map of the district, showing A, B, C, and D on a scale, say, 1 foot to a mile. We start with A, and draw the lines A to B, B to C, C to D, in the proper directions, and to scale. This fixes D. On the map we can draw the line AD, and measure the angles which it makes with AB and CD. We now know in what directions to bore at A and D.

This method shows that the problem can be solved by a common-sense method, though not very accurately. Since we are working to the scale of 1 foot to a mile, an error of $\frac{1}{100}$ inch in our drawing will lead to an error of $1\frac{1}{2}$ yards in the actual work. And in drawing the figure we may easily make several mistakes of $\frac{1}{100}$ inch. Drawing a figure, then, is not sufficient to give a really good answer, but it gives us a general idea of what is needed. First attempts often turn out like this: they give us the germ of an idea: we have to work at this idea until it becomes practical. A mathematical invention goes through the same stages as a mechanical one: first an idea, then a toy, then a commercial proposition.

Trigonometry represents an attempt to improve on the method of drawing. The argument runs rather on the following lines. By drawing the map on a larger scale, we could get a more accurate answer to the tunnel problem. There seems to be no limit to the accuracy we could get by taking our plan large enough. Given the lengths and directions of the lines AB, BC, CD to a high degree of accuracy, we could find (by drawing on an immense scale) the length and direction of AD to a high degree of accuracy. It seems likely that some rule connects the answer with the facts given. We could collect information on the problem, taking A, B, C, and D in different positions, and trying to see how the length of CD, and its direction, depended on the other measurements given. The aim would be to find the rule: once we had this rule, we could work out AD to as many decimal places as we chose, without doing any drawing at all.

In trigonometry, then, we consider problems which could be solved by drawing, problems which therefore possess a definite answer (it is a waste of time to try any problem by trigonometry if you are not given sufficient facts to solve it by drawing: trigonometry is not magic): we try to discover *what rule* gives this answer, so that we may be able to find the answer by a formula, instead of by drawing. The aim is therefore to replace drawing by calculation.

Such a question can be tackled, in the first place, only by experiment. Lengths, directions – these are real things. They will not take orders from us: they follow the laws of their own nature. We can find what they do only by *observing* them.

But, of course, we shall not begin by experimenting with such a complicated problem as that of the tunnel and the four points *A*, *B*, *C*, *D*. It is rarely wise to attack a problem of a *new type* directly. It is better to make up a much simpler problem of the same type, experiment with that, and see if the method which solves the simple problem throws any light on the complicated one. In map-making, the simplest problems deal with three points only (hence the name trigonometry, i.e., three-line-ology). In particular, triangles containing a right-angle are easy to study.

The Measurement of Angles

It is easy enough to measure the length of a line. It is not so obvious how an angle is to be measured. Two methods are used.

The first method, measurement in degrees, has something in common with the markings on a clock-face. On a clock the numbers 1 to 12 are evenly spaced out, around the rim of the clock. If the hour-hand goes from 12 to 3, we know it has gone a quarter of the way round. To obtain degrees, the circle has to be divided, not into 12, but into 360 equal parts. Each part is called a degree. There is no deep reason for choosing the number 360. Turning through one quarter of the circle (a right angle) corresponds to 90 degrees, usually shortened to 90°. From 12 to 1 on the clock is 30°.

The second method is known as *radian* measure, and is particularly suitable for questions connected with speed. It may be explained as follows. Suppose we have a wheel, 1 foot in radius, fixed to an axle. A string passes round the rim of the wheel, one end being fixed to the wheel, rather like the rope on a capstan or a crane. By pulling the string, we may cause the wheel to revolve. It is clear that we can measure how much the wheel has turned by measuring the amount of string unrolled. When 1 foot of string has been unrolled, the wheel is said to have turned through *one radian*. When x feet have been unrolled, the wheel has turned through x radians.

It is easy to measure an angle in radians. We take a piece of wood, made in the shape of a circle of 1 foot radius. To measure a given angle, we place the centre of the circle. O, at the point of the angle, and mark the points A and B where the lines cross the rim (Fig. 16). We then wind a tape measure *around the rim* (*not* straight) and measure the distance from A to B. If this distance is $\frac{2}{3}$ foot, the angle is $\frac{2}{3}$ of a radian. If we are told that the hand of a clock turns through 10 radians, we measure 10 feet round the rim. We shall, of course, complete more than one turn: where we end up gives the angle 10 radians. If a wheel, of radius 1 foot, with a fixed centre, is turning at the rate 1 radian a second, any point

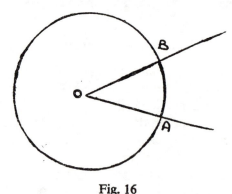

Fig. 16

on the rim is moving at 1 foot a second. Radians are therefore convenient for mechanical problems about ropes being wound on to wheels, for wheels rolling on the ground, and generally for theoretical purposes. If, anywhere in a book on mathematics, you see any statement about 'an angle x' or 'the angle $3\cdot5$', with

nothing more said, you must expect this to mean '*x radians*' or
'3·5 *radians*', *not x* degrees or 3·5 degrees. 3·5 degrees is always
written as 3·5°. If nothing is said about degrees, the angle is in
radian measure. For mathematicians radian measure is the most
natural to use, since it gives the simplest results.

If you measure right round the rim of a circle with the radius
1 foot, you will find the length to be (roughly) 6·28 feet. So one
complete turn, 360°, is the same thing as 6·28 radians. This
number, 6·28, is not a pleasant number to meet, but we cannot do
anything about it. The universe is so made that this number turns
up: it is not the fault of mathematicians. We cannot get away
from 6·28. If we measure in degrees, so as to make one complete
turn a convenient number, 360°, we find that the complication
turns up elsewhere. When a wheel turns through 360° a second,
the speed of points on the rim (supposing the radius, as before,
to be 1 foot) will be 6·28 feet a second. Accordingly, *radians* are
used for most questions connected with speed, or with the rims
of circles: degrees may be used when we are measuring the angles
of things which do not move – fields, for instance.

If you are not familiar with radian measure, you may find it
worth while to cut out a large circle, and to mark the two scales
around the rim – degrees and radians. Whenever you meet an
expression such as 202° or 2·78 radians, you will be able to look
at your circle and see what angles these expressions represent. It
will be best if you put 0° at the position '3 o'clock', and go anti-
clockwise, so that 90° will come at the top (12 o'clock), 180° at
9 o'clock, 270° at 6 o'clock, and 360° back again at 3 o'clock.
0 radians will also be at 3 o'clock, 1·57 at 12 o'clock, 3·14 at 9
o'clock, 4·71 at 6 o'clock, 6·28 at 3 o'clock again. It has become
a custom among mathematicians to think of angles in these
positions (I do not know why), and it will save misunderstanding
if you do the same.

Sines and Cosines

We can now proceed to our experiments on right-angled triangles.
It is again to be emphasized that the beginnings of the subject

must be experimental. I cannot imagine anyone making any progress who simply sat looking at a right-angled triangle, hoping to be inspired with a method of reasoning out the problem. We must begin with experiments, and then see how much these help us.

Problem: a railway line makes an angle of 5° with the level and is perfectly straight: if a train travels 10,000 feet along the line, how many feet does it rise? No use thinking about it – let us measure and see. We find the answer, correct to the nearest tenth of a foot, to be 871·6 feet (you must take my word for this, unless you are prepared to carry out the experiment for yourself). Nothing particularly simple about the answer: it does not suggest any way of calculating the result without measurement.

But this result does one important job for us: it means that we need not do any more measurements of this particular type, on railways rising at 5°. If we are asked, 'How much does the train rise if it travels 100 feet?' we know the answer immediately. Since the line is straight, the train climbs steadily. In 10,000 feet it will climb 100 times as much as in 100 feet. Therefore, in 100 feet of travel it rises 8·716 feet. In fact, for each foot the train travels, it rises 0·08716 feet (correct to five places of decimals). If it travels x feet, it rises $0·08716x$ feet.

In the same way, to any angle (measured in radians or degrees) there corresponds a number. In travelling x feet along a slope of 13° we rise $0·22495x$ feet: for 30° the formula is $0·50000x$. (Note this, our first simple result, $\frac{1}{2}x$ corresponding to 30°.) It is convenient to have a short way of referring to the numbers that arise in this way. We therefore give them a name, *sines*. (The name goes back to the time when learned men of all countries wrote to each other in Latin: it means a 'bowstring' – the reason for this name may be guessed from Fig. 17.) We say that 0·08716 is the sine of 5° (usually shortened to *sin* 5°), that *sin* 13° = 0·22495 and *sin* 30° = 0·5. 30° is 0·52360 radians, 13° is 0·22689 radians, 5° is 0·08727 radians. (To get these results so accurately we should have to use a circle of 10,000 feet, and measure round the rim). So we may also write *sin* 0·53260 = 0·5, *sin* 0·22689 = 0·22495, *sin* 0·08727 = 0·08716.

Note, in passing, a fact which leaps to the eye: in radian measure, though not in degree measure, the sine of an angle and the angle itself, for fairly small angles, are nearly the same number. The smaller the angle, the nearer it is to its sine. 0·5 is *sin* 0·53260; the two numbers 0·5 and 0·53260 differ by 0·0326. But *sin* 0·08727 is 0·08716. The two numbers here, 0·08727 and 0·08716, differ only by 0·00011. (This fact we have discovered without any effort: you will usually find that, as soon as you start to collect evidence, the discoveries make themselves.) This result suggests that there is some simple law connecting an angle, in radian measure, and its sine. We shall not be surprised when, in Chapter 14, we find a series giving sin x in terms of x. It is important to remember that this series holds *only when the angle is measured in radians, not for degrees*. (Look for the series in Chapter 14, and write a note there, in the margin, to this effect.)

The *cosine* of an angle is defined in a similar way. If an aeroplane starts from an aerodrome and flies 10,000 feet in a straight line at 30° to the level, we know how high it is. It is 10,000 *sin* 30° feet above the ground. Directly underneath the aeroplane there is a certain point on the ground. How far is this point from the aerodrome? By measurement is it found to be 8660·3 feet. Every extra foot the aeroplane flies (still keeping in the original line), this point moves 0·86603 feet further from the aerodrome. If the plane flies x feet, the point moves 0·86603x feet. We call 0·86603 the cosine of 30°, and write, for short, 0·86603 = *cos* 30° = *cos* 0·52360. (The last figure gives 30° in radian measure.)

In short: if we move on a straight line, making any angle t with the level, each foot we travel increases our height in feet by a certain number, called *sin t*, and carries us sideways through a certain number of feet, the number being called *cos t*.

We can easily make a model to demonstrate the meaning of *sin t* and *cos t*. Draw a circle of 1 foot radius, mark around it scales both for degrees and radians. Pin it to a wall or blackboard. Take a strip of cardboard, something over a foot in length. Fasten one end of it by a tin-tack or drawing-pin through the centre of the circle, so that the strip is free to turn. One foot from the centre pierce a small hole in the strip, and hang a plumb-line from this

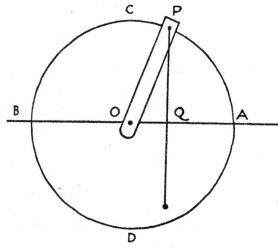

Fig. 17

hole. The arrangement is sketched in Fig. 17. The line *BOA* is level. The thread hanging from the small hole, *P*, crosses *BOA* at the point *Q*. *P* is higher than the line *AOB* by a height *PQ*, and it is a distance *OQ* to the right of *O*. As *OP* is 1 foot in length, the distance *PQ*, measured in feet, is equal to the sine of the angle *AOP*, and the distance *OQ*, also in feet, is equal to the cosine of *AOP*. To make a rough table of sines and cosines it would be better to have *OP* 1 metre long, then, by measuring *OQ* and *PQ* to the nearest millimetre, we should find results certainly true to two places of decimals, perhaps to three.

Actually doing this, or seeing it done, is helpful to the type of person who is good at games but has not a vivid imagination.

We have now explained what sines and cosines are. If you are given any statement about them, you can test for yourself whether it is true or not.

One point is worth mentioning. We have said that *sin t* represents the height of *P* above the line *BOA*. But if *P* is at the angle 270°, which is the same as 4·71 radians, *P* is 1 foot *below BOA*. Being 1 foot below *BOA* we may call being —1 foot above *BOA*. We therefore say that *sin* 270°, or *sin* 4·71, is —1. Similarly, the sines of all the angles between 180° and 360°, between 3·14 and

6·28 radians, have *minus* signs. Also, we take *cos t* to mean the distance Q is to the *right* of O. If Q lies to the left of O, as it does for angles between 90° and 270°, between 1·57 and 4·71 radians, the cosine has a *minus* sign.

Check for yourself the following table.

ANGLE (degrees)	0	90	180	270	360
(radians)	0	1·57	3·14	4·71	6·28
SINE	0	+1	0	—1	0
COSINE.. ..	+1	0	—1	0	+1

An explorer recording his journey will observe in what direction he travels, and for how many miles. If we regard East as corresponding to 0°, North will be 90°, and so on. On a map, North usually means *upwards*, East means *to the right*, so it is easy to apply our railway and aeroplane illustrations to map-making. If an explorer goes 100 miles in the direction 20°, then 50 miles in the direction 40°, what is his new position? 100 miles in the direction 20° is the same as going 100 *cos* 20° East, and then 100 *sin* 20° North. 50 miles in the direction 40° is the same as 50 *cos* 40° East, then 50 *sin* 40° North. If we have tables, we can work out these numbers, and then it is simple to add up, and find the total distance he had gone to the East, and the total distance to the North. Two things to note : (*a*) this method applies to the tunnel problem, Fig. 15, (*b*) it throws some light on the *minus* signs mentioned above. If the explorer, after doing the journey just mentioned, goes a further 30 miles in the direction 135° (i.e., North-West), this *increases* his distance to the North (*sin* 135° is +), but *decreases* his distance towards the East (*cos* 135° is —). In fact, by using the + and — signs in the definition of sine and cosine, we save ourselves the need for any further thought : we just have to write down *distance gone* times *sine*, for each part of the journey. The signs + and —, which then appear, automatically show whether the numbers have to be added on, or taken away.

The Formulae of Trigonometry

Certain other terms, besides sine and cosine, occur in trigonometry – namely, tangent, cotangent, secant, and cosecant.

These are, however, merely abbreviations, and do not bring in any essentially new idea: the subject could be mastered without using these terms at all. We shall therefore not deal with them here, but proceed to study the properties of the sines and cosines.

We shall, of course, try to discover properties of sines and cosines which are useful for our purposes. We have two particular problems in mind, and one rather general use as well.

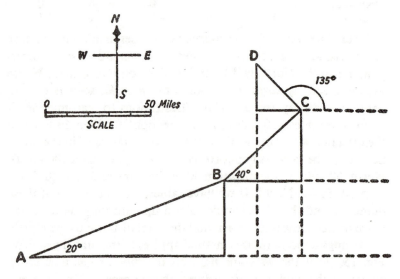

THE EXPLORER'S JOURNEY

The explorer travels from *A* to *B*, from *B* to *C*, and then from *C* to *D*. *AB* is 100 miles, *BC* is 50 miles, *CD* is 30 miles. He records each part of his journey, and works out (by the method explained in the text) how far each part carries him towards the East, how far towards the North. Distances West and South appear with *minus* signs, since 10 miles farther West means 10 miles *less* to the East. The record appears as below:

	Distance.	Direction.	To East.	To North.
A to B.	100 miles	20°	94·0 miles	34·2 miles
B to C.	50 ,,	40°	38·3 ,,	32·1 ,,
C to D.	30 ,,	135°	—21·2 ,,	21·2 ,,
Whole journey A to D.			111·1 miles	87·5 miles

The first problem assumes that we have already in our possess-
ion satisfactory tables of sines and cosines, and is known as the
solution of triangles. It is a problem which naturally arises in
surveying. We are given certain information about a triangle,
sufficient to enable us to draw the triangle, and are asked to find
the remaining quantities. For instance, in any triangle ABC we
might be told the length of AB, and the angles ABC and BAC, and
asked to find the lengths AC and BC. This problem frequently
arises in map-making, in the construction of range-finders, in
determining the position of a ship at sea by taking the bearings of
two lighthouses, in locating submarines, etc.

Surveyors and seamen are in a position to buy printed books
of tables, containing sines and cosines and other information.
But someone first had *to make these tables* – this is our second
problem – and several of the properties of sines and cosines were
discovered with this object in view. The interest which mathematicians of the sixteenth century showed in algebra was partly
due to the fact that equations had to be solved before trigono-
metric tables could be made. *

Thirdly, it is desirable to know the properties of sines and
cosines on quite general grounds. They arise in many problems,
and the work can often be made shorter and simpler if the formulae are known. An example of this will be given later.

Pythagoras' Theorem

In Fig. 17 the sides OQ and QP have the lengths *cos t* and *sin t*,
where t is short for the angle QOP. Students usually find this
figure easy enough to grasp, but they do not always recognize it
when it occurs in an unusual position, or on a different scale.
For instance, in Fig. 18, DF makes an angle t with DE, and DF
has the length 1. The line EG is drawn at right angles to DF. It
is clear enough that the triangle DEF has the same shape as the
triangle OQP. It may not be so obvious that there are two other

* See Zeuthen, *History of Mathematics in the Sixteenth and Seventeenth Centuries*, Chapter II, section 4. I am not sure if this work can be
had in English.

triangles in the figure with the same shape. But this is so. If you cut out pieces of paper just large enough to cover the triangles *DEG* and *FEG*, you will find that it is possible (after turning the paper over) to lay these triangles in the positions *LVW* and *LTU*. The triangle *LMN* is exactly the same size and shape as *DEF*. It is now obvious that the three triangles have the same shape, and differ only in size.

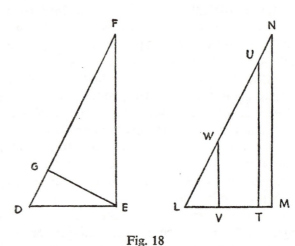

Fig. 18

How long are the lines *DG* and *GF*? The line *DG* can be put in the position *LV*, and is therefore *cos t* times *LW*. But *LW* has the same length as *DE*, which is equal to *cos t*. Accordingly, *DG* must be *cos t* times *cos t*, or $(cos\ t)^2$. In exactly the same way, it may be shown that *GF* has the length $(sin\ t)^2$. But *DG* and *GF* together make up *DF*, which we drew with the length 1 when we began. It follows that:

$$(cos\ t)^2 + (sin\ t)^2 = 1.$$

In Chapter 2 we met a triangle with the sides 3, 4, and 5, and a right-angle between the sides 3 and 4. If we draw this triangle on a scale one-fifth as large, we shall have sides $\frac{3}{5}$, $\frac{4}{5}$, and 1. For this triangle, then, $DE = \frac{3}{5}$ and $EF = \frac{4}{5}$, and $cos\ t = \frac{3}{5}$, $sin\ t = \frac{4}{5}$. The formula given above thus becomes:

$$(\tfrac{3}{5})^2 + (\tfrac{4}{5})^2 = 1, \text{ or } 3^2 + 4^2 = 5^2.$$

It is because of this relation between 3, 4, and 5 that the triangle is right-angled. Another such triangle is 5, 12, 13. $5^2 + 12^2 = 13^2$. If we draw an angle whose cosine is $\frac{5}{13}$, its sine will be $\frac{12}{13}$.

Here we have the answer to the question raised in Chapter 2: the proof given above is, in essentials, that given in Euclid. The result is usually known as Pythagoras' Theorem, and can be stated: if a, b, c are the lengths of the sides of a right-angled triangle, then $a^2 + b^2 = c^2$. This result is essentially the same as the result we have just found. For if t is the angle between the sides a and c, $a = c \cos t$ and $b = c \sin t$. (This result is obtained by enlarging the scale of our standard triangle, OPQ, c times.) So $a^2 + b^2 = (c \cos t)^2 + (c \sin t)^2$. This last expression is equal to c^2 multiplied by $(cos\ t)^2 + (sin\ t)^2$. By the result proved above, this is the same as c^2 multiplied by 1: that is, c^2. So $a^2 + b^2 = c^2$ follows from our earlier result, by simple algebra.

The Cosine Formula

So far we have considered only triangles containing a right-angle. We will now consider a more general problem. Suppose we have a triangle ABC, and we know the lengths AB and AC, and the size of the angle BAC. How long is BC? (It might be impossible to measure BC directly, owing to mountains, rivers, swamps, etc.)

In books on trigonometry it is usual to write a, b, c for the lengths of the sides BC, CA, AB and to write A, B, C for the three angles of the triangle. Thus, a is the side opposite the angle A, etc. Our problem is: given b, c, A, to find a.

Can this problem be solved at all? Are the facts given sufficient to allow us to draw a plan of ABC? They are. The problem can be solved by drawing: it is a reasonable problem to try.

Can we solve it with the help of tables of sines and cosines without drawing? What are the tables of sines and cosines? They are the result of experiments made on right-angled triangles. Sines and cosines therefore tell us nothing about a figure, unless that figure can be split up into right-angled triangles. Can we split ABC up into right-angled triangles? Very easily indeed. All we have to do is to draw CD at right-angles to AB (Fig. 19). We

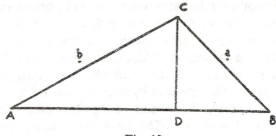

Fig. 19

now have two right-angled triangles, ADC and BDC. What do we know about these?

Triangle BDC: not much hope here. We want to find BC, but all we seem to know is that BDC is a right-angle.

Triangle ADC: quite a different story. We know $AC = b$, and we know the angle $CAD = A$. In fact, we know everything about this triangle: we have exactly the same information as we had in the railway problem, when we were told the angle the railway made with the level (A) and the distance the train travelled (b). The height CD is therefore $b \sin A$, and the distance sideways, AD, is $b \cos A$.

This new information helps us with triangle BDC. It tells us the length CD, and shows how DB can be found. For $AB = c$, and $AD = b \cos A$. DB is what is left when AD is taken away from AB. So DB must be equal to $c - b \cos A$.

We now know enough about the triangle BDC to fix it completely. We know DC, BD, and the angle CDB is a right-angle. BC can be found by Pythagoras' Theorem, for $BC^2 = DC^2 + DB^2$. Writing for BC, DC, and DB the lengths found for them, we have:

$$a^2 = (b \sin A)^2 + (c - b \cos A)^2.$$

This formula can be put in a more simple form. Before doing this, we may glance for a moment at the strategy by which we reached this point. The difficult thing in a mathematical problem is *to get started*. Before writing down any calculations at all, one should always prepare a plan of campaign. Otherwise one wanders around like a rudderless ship. While making this plan, forget all the difficulties that may come in the actual calculations. Try simply to build a framework connecting what we know with what we want to know. It is sometimes useful to draw a pencil figure, and to mark with ink those lines whose lengths are given, or angles whose size is known. Then mark with ink lines and angles which can be calculated from those already marked. And so go on, keeping a record of the steps.

For the present problem our plan would be as below.

> Line AC and angle DAC given. (Ink these in.)
>
> AD and DC can be calculated. (Ink these.)
>
> AB is given. (Draw a line in ink, just below AB, so as not to blot out the line AD already drawn.)
>
> So DB is found from AB minus AD.
>
> BC is found from DC and DB by Pythagoras.

Do not worry if you have forgotten the formula $AD = b \cos A$, or the exact result of Pythagoras' Theorem. All you need to know in making this plan is that *a formula exists*: that the thing *can* be worked out. In real life (which is more important than examinations) you can always carry a book of formulae with you, and look them up. But no book will tell you on what lines to tackle a problem: that you must learn for yourself, by practice.

Now let us return to the formula we found for a^2. By simple algebra, we can work it out and obtain:

$$a^2 = b^2 \sin^2 A + c^2 - 2bc \cos A + b^2 \cos^2 A.$$

$\sin^2 A$ is the usual way of writing what, until now, we have written $(\sin A)^2$, and $\cos^2 A$ means the same as $(\cos A)^2$. Writing in this way saves a lot of brackets.

We notice that b^2 comes twice in this result. First we have b^2 multiplied by $\sin^2 A$, then b^2 multiplied by $\cos^2 A$. The total amount of b^2 that appears is therefore $\sin^2 A + \cos^2 A$, which is equal to 1. It follows that:

$$a^2 = b^2 + c^2 - 2bc \cos A$$

which is the usual formula given in text-books, and used in problems.

This is an example of the way in which formulae can be shortened by using the properties of sines and cosines. Earlier we promised that such an example would be given.

The Addition Formulae

Now, for some results which arise naturally in connexion with the problem of making tables, though they are also useful for general knowledge.

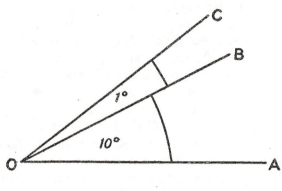

Fig. 20

Suppose that we are setting out to make a very accurate table of sines and cosines, and that (at great labour and expense) we have constructed large triangles and found accurate values for *sin* 1°, *cos* 1°, *sin* 10°, and *cos* 10°. It would be possible to keep on making fresh triangles, and to find by measurement *sin* 11°, *sin* 12°, etc. If carried out on a really large scale, this work would be very troublesome. It would be natural to think, 11° is 10° + 1°. Is it possible to use this fact in some way, and to find *sin* 11° by calculation, from what we know about 10° and 1°? If we can do this it will be very convenient, for the same method will give us information about 12°, since 12° = 11° + 1°, and we may continue thus as long as we care to do.

Our problem is : we have found, by measurement, that *sin* 1° = 0·01745, *cos* 1° = 0·99985, *sin* 10° = 0·17365, *cos* 10° = 0·98481. What are *sin* 11° and *cos* 11°?

The main difficulty in this problem is to draw a figure that brings out the facts clearly. It is easy enough to draw an angle of 10° with a further angle of 1° on top of it, as in Fig. 20. This illustrates the fact that 11° = 10° + 1° all right. But it does not

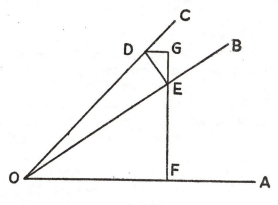

Fig. 21

tell us much about 10° and 1°. We have to take it on faith that
the angles marked 10° and 1° *are* in fact 10° and 1°. There is
nothing in the figure to show that they are: in particular, there is
nothing to link them up with *sin* 1°, *sin* 10°, etc. (Actually, for the
sake of clearness in the figure, it is necessary to draw the angles
rather larger than they actually are.)

We want to bring out the fact that *BOC* is the angle, 1°, whose
sine is 0·01745, and whose cosine is 0·99985. To do this we must
bring in a right-angled triangle. Take *D* at a distance 1 from *O*,
and draw *DE* at right angles to *OB* (Fig. 21). Then we know
OE = *cos* 1° = 0·99985, and *DE* = *sin* 1° = 0·01745.

How is *sin* 11° to be brought into the picture? *OD* has the length
1 and makes the angle 11° with *OA*. The height of *D* above *OA* is
therefore *sin* 11°. It is this height that we wish to find.

But this is easy to do: it is exactly the same problem as we had
when the explorer went 100 miles in one direction, and then 50
miles in another. We can get from *O* to *D* by going first from *O*
to *E*, then from *E* to *D*. We know the length and direction both
of *OE* and of *ED*.

Draw an upright line *FEG* through *E*, *F* is on *OA*, and *G* is a
point at the same height as *D*, so that *FG* equals the height of *D*
above *OA*, that is *FG* = *sin* 11°.

As *FG* = *FE* + *EG*, the problem is solved as soon as we can calculate *FE* and *EG*. *FE* presents no difficulties. $OE = 0.99985$ and *OE* makes an angle 10° with *OA*, so

the height $FE = 0.99985 \; sin \; 10° = 0.99985 \times 0.17365$.

EG can be found from the triangle *EGD*, which has a right-angle at *G*. The triangle *GED* could be obtained by turning the triangle *EOF* through a right-angle, and then making it shrink to a smaller scale. The angle *DEG* is, in fact, the same as the angle *EOF* – that is, 10°. Accordingly, $EG = ED \; cos \; 10° = 0.01745 \times 0.98481$. Adding these two results together, we obtain the length of *FG* – that is, *sin* 11°.

The result we have just found may be written:

$$sin \; 11° = cos \; 1° \; sin \; 10° + sin \; 1° \; cos \; 10°.$$

There is nothing particular about the numbers 1 and 10. The same argument could be carried through for any two numbers, *x* and *y*, and we should find:

$$sin \; (x+y)° = cos \; x° \; sin \; y° + sin \; x° \; cos \; y°.$$

You should find no difficulty in working out *cos* 11°, the distance that *D* is to the right of *O*, and the corresponding general formula for $cos \; (x+y)°$.

Other Formulae

The formulae we have considered must be regarded as samples. There are other formulae in trigonometry, which, for the most part, can be found by arguments very similar to those given above. Some books contain vast masses of results. For most purposes, a few formulae and a few straightforward methods are quite sufficient. If you are studying trigonometry for some definite purpose – e.g., surveying, or navigation – you will do well to obtain a book on that subject, and see what formulae of trigonometry are actually used, and for what problems.

Differentiating Sines and Cosines

It often happens that sines and cosines occur in problems about the movement of machinery, the vibrations of some object, or the changes in electric currents. All these, being problems of change of speed, call for differentiation. It is therefore worth while to study the question: how fast do *sin t* and *cos t* change when *t* changes?

We shall study this problem by means of the model illustrated in Fig. 17. We suppose the point *P* begins at *A*, and travels round the circle at a steady speed of 1 foot per second. After *t* seconds it will have travelled *t* feet, and the angle *AOP* will therefore be *t* radians. (The results we shall obtain hold only when the angle is measured in radians.)

We know that *sin t* measures the height of *P* above *AOB* after *t* seconds. This we call, for short, *y* feet, so $y = sin\ t$. *cos t* measures the distance *P* lies to the right of *O* after *t* seconds. This we call *x* feet, so $x = cos\ t$. Of course, if *P* lies below *AOB*, *y* will be a number with a minus sign: *x* will be minus if *P* lies to the left of *O*. In the illustration, *x* equals the length of *OQ* in feet, *y* equals the length of *PQ* in feet.

Note that these signs, *x* and *y* have *no connexion at all with any signs x and y that may have been used in other chapters.* For instance, in Chapter 10 *x* stood for the number of seconds that had passed, and in Chapters 11 and 12 we discussed the expression $\frac{dy}{dx}$. In this section *t* is the sign used for 'the number of seconds': *x* and *y* have simply the meanings given to them in the last paragraph.

The speeds with which *x* and *y* change will be $\frac{dx}{dt}$ and $\frac{dy}{dt}$. We shall also write these as *x'* and *y'*. So *x'* is to mean the speed at which the length *OQ* changes, and *y'* the speed at which the length *PQ* changes. (If *P* lies below *AOB*, we shall have to continue the line of the plumb-line upwards, until it crosses *AOB*. This gives *Q*.) We have already explained carefully what is meant by speed, and how speed can be measured. The meaning of *x'* and *y'* should be clear.

There are four points on the circle at which it is particularly easy to see what is happening. These are the highest point, *C*, the lowest point, *D*, together with the two points, *A* and *B*. At *C* and *D* the track of *P* is level, at *A* and *B* it is upright.

As the track is level at C and D, the height of P cannot increase or decrease as it passes these points. As y' measures the speed with which the height of P changes, it follows that y' must be nothing when P passes C or D. It may be easier to see this result if you consider that P is moving upwards just before it reaches C (so y' is $+$), downwards just after it passes C (so y' is $-$). At C, y' is just at the moment of changing from $+$ to $-$, and must be 0. (Compare the remarks in Chapter 11, on the meaning of y'.)

The same argument shows that $x' = 0$ when P is at A or B.

What is x' when P is at C? At C the curve is level. Just for an instant the point P is neither rising nor falling, but is simply travelling towards the left, with a speed of 1 foot per second. In other words, at this instant x is decreasing at the rate 1 per second. That is, $x' = -1$.

At the point D, P is moving towards the right, with a speed of 1 foot a second. So $x' = 1$. In the same way, we may find y' for the points A and B. At A, P moves upwards, y is increasing, $y'=1$. At B, P is moving downward, $y' = -1$.

The points A, C, B, and D are reached after 0, 1·57, 3·14, 4·71 seconds (roughly). We may extend the table given earlier in this chapter, thus:

POSITION	A	C	B	D
$t=$time in seconds$=$angle in radians	0	1·57	3·14	4·71
$x=\cos t=OQ$	$+1$	0	-1	0
$y=\sin t=QP$	0	$+1$	0	-1
x'	0	-1	0	$+1$
y'	$+1$	0	-1	0

This table suggests something. The row y' is the same as the row x: the row x', except for a change of sign, is the same as y. This *suggests* that $y' = x$, and $x' = -y$. These results are, of course, not proved: they are not even made likely. We have taken only four points of the entire circle as evidence. (The law $y' = x^3$ would fit the table equally well.) But, as a bold guess, we might at least *investigate* the results.

It is left to the reader to prepare further evidence, using the points between A and C. A printed table of sines and cosines may be used. Remember to record the angle, t, in *radians*. (One degree $= 0 \cdot 01745$ radian.) In this way you can convince yourself that the guess was actually correct.

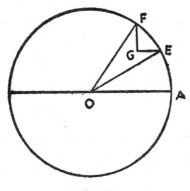

Fig. 22

It is also possible to see this result from a figure. In Fig. 22, E represents the position of P after any time, t seconds. That is, t feet of tape, wound round the circle, starting from A, would finish at E. A little later, P will have moved to a point, F, a little farther round the circle. The extra piece of tape, EF, will have the length Δt feet. If F is very close to E, the tape EF will be very nearly straight, and we shall not be making a serious error if we think of Δt as giving the length of the *straight line EF*.

The line EG is level, and the line GF is upright. So GF represents the increase in the height of P, as P goes from E to F – that is, $GF = \Delta y$. The angle GFE, you will find, is very nearly equal to the angle AOE, which is t radians. Accordingly, $GF = FE \cos t$, very nearly. (The smaller EF is, the nearer this equation comes to the truth.) That is, $\Delta y = \Delta t \cos t$, very nearly. When Δt becomes smaller and smaller, we find $\dfrac{dy}{dt} = \cos t$.

In the same way, GE is equal to the *decrease* in x, $-\Delta x$, and $GE = EF \sin t$, very nearly, which leads to the result $\dfrac{dx}{dt} = -\sin t$.

As x stands for $\cos t$, and y stands for $\sin t$, we may write these results as follows:

$$\frac{d\,(sin\,t)}{dt} = cos\,t \qquad\qquad \frac{d\,(cos\,t)}{dt} = -sin\,t.$$

This is shorter than saying, 'If $y = sin\,t$, $\frac{dy}{dt} = cos\,t$, etc.', and it means the same thing. We shall quote the result in this form in Chapter 14, when we find series for $cos\,t$ and $sin\,t$, or, at any rate, for sine and cosine. I cannot promise that we shall always use the letter t when sine or cosine occurs.

Movement in a Circle

We saw in Chapter 10 that the force acting on a moving weight can be found if we know mx'' and my''. It often happens in machinery that a heavy weight goes round in a circle, as, for instance, any part of a flywheel, or the metal attached to a loco-motive wheel (though this also travels along the line). An aero-plane looping the loop or a motor-car going round a corner raises a similar problem.

We may therefore consider a weight attached to the point P of Fig. 17, and see what force would be required, to make it move in the desired way. Since $y = sin\,t$, $y' = cos\,t$, and y'' (the rate at which y' changes) is therefore $-sin\,t$. In the same manner, we ﬁnd $x'' = -cos\,t$. There is no difficulty in finding x'' and y'', and the total force acting on the weight at P can be found easily by anyone who has had some practice in the elementary prob-lems of statics and dynamics.

EXERCISES

1. Cut out a circular disc of cardboard, and mark a scale for measuring angles in radians around the edges, by the method ex-plained in this chapter. On the same disc mark a scale for measuring angles in degrees.

Draw on a piece of paper an angle of ⅜ radian, 1 radian, 2½ radians, 5 radians, 10 radians.

How many radians are 10°, 50°, 95°, 184°?

2. Make an actual model from the design of Fig. 17. From this

make a table giving the sines and cosines of 5°, 10°, 15°, etc. (as far as 90°), to two places of decimals. Check your results for sines from printed tables.

3. Write down, from your results for question 2, *sin* 10°, *sin* 20°, etc. up to *sin* 80°. Write down the cosines in the opposite order: *cos* 80°, *cos* 70° ... *cos* 10°. What do you notice about the two lists? What can you say about *sin* $x°$ and *cos* $(90—x)°$? Can you see any reason for your result?

4. From your model (question 2) find to two places of decimals *sin* 100°, *sin* 110° ... *sin* 170°. Compare these with *sin* 10°, *sin* 20° ... *sin* 80°. What do you notice about the two sets? What formula connects *sin* $(180—x)°$, and *sin* $x°$?

5. Find from your model *cos* 100°, *cos* 110° ... *cos* 170°. (For all of these Q is to the left of O, so the cosines all have a minus sign.) Compare these with *cos* 10°, *cos* 20° ... *cos* 80°. What formula connects *cos* $(180—x)°$ and *cos* $x°$?

Also compare the set just found, *cos* 100° ... *cos* 170° with *sin* 10° ... *sin* 80°. Is there a formula connecting *cos* $(90+x)°$ with *sin* $x°$? If so, what is it?

6. The printed tables give the sines of angles between 0° and 90°. To find the sines of angles between 90° and 180°, and to find cosines, we have to use the results of questions 3, 4, and 5. For example, the tables tell us *sin* 37° $= 0·6018$. What are the values of *cos* 53°, *sin* 143°, *cos* 127°?

7. An airman flies 200 miles in the direction 37° North of East. How many miles to the East, and how many miles to the North does his position change? (*Note* – For long flights it is necessary to take account of the fact that the earth is shaped like a ball. All questions in this chapter refer to short journeys, for which the earth may be supposed flat.)

8. Find how many miles East and how many miles North of A is the point C given by the explorer's log below:

A to B 30 miles, in the direction 40° N of E.

B to C 10 miles, due West.

9. Do the same for the journey below:

A to B 40 miles, in direction 70°.

B to C 20 miles, in direction 110°.

10. Also for:

A to *B* 100 miles in direction 315° (i.e., South-East).

B to *C* 150 miles in direction 80°.

11. An aeroplane has to fly to a town 100 miles away. By mistake it flies in a direction which is 2° off the correct course. When it has flown 100 miles, how far will it be from the town?

12. Ipswich is 65 miles from London, in the direction 36°. Peterborough is 75 miles from London, in the direction 95°. Birmingham is 105 miles from London, in the direction 139°.

How far is Ipswich from Peterborough, Peterborough from Birmingham, and Birmingham from Ipswich?

(This can be done by means of the formula $a^2 = b^2 + c^2 - 2bc . \cos A$. To find the distance from Ipswich to Peterborough, we may take London as *A*, Ipswich as *B*, Peterborough as *C*. A similar calculation gives the other two distances. Remember that $\cos 103°$, which comes in the formula for the distance from Birmingham to Ipswich, has a minus sign. Check your calculations by drawing.)

CHAPTER 14

ON BACKGROUNDS

'*Recent workers in the sociology of science have stressed that experimental science arose from the theorists' taking account of the crafts. On the other hand, the crafts have often failed to learn from the theorists almost down to our own day.*' – H. T. Pledge, *Science Since 1500.*

STUDENTS of mathematics often have the experience of understanding the proof of some results, but not being able to see what it is all about. The subject remains 'up in the air', a disconnected piece of knowledge. As memory depends on connexions, the result is hard to remember. In ordinary life we remember familiar objects well, because other things continually remind us of them, and thus refresh their images in our mind. Students rightly feel uneasy when they are asked to remember something unconnected with the rest of life: the mind cannot work efficiently unless it is properly treated.

In elementary algebra this 'unearthly' feeling is easily aroused. Many text-books, for instance, explain quite accurately what is meant by an arithmetical progression, or a geometrical progression. The teacher (who may be passionately interested in some other subject, and forced to teach mathematics without understanding it) follows the text-book and teaches A.P.s and G.P.s simply because they are in the book.

We have already (without noticing it) had two examples of arithmetical progressions. The man falling off a house travels 1 foot in the first quarter-second, 3 feet in the next quarter-second, 5 feet in the third quarter-second, 7 feet in the fourth, and so on. The total distance gone in one second is $1+3+5+7$ feet. In the set of numbers, 1, 3, 5, 7, etc., each number is 2 more than the previous one. A set of numbers in which each number is bigger (or smaller) than the number before it by a fixed amount is called an arithmetical progression (A.P.).

The second example was in Chapter 12, when we added together the numbers, 0, 0·01, 0·02, etc., up to 0·09. These numbers also form an A.P. Later in the same section we saw that we could have found a more accurate estimate of $\int_0^1 x \, dx$ if we had divided

the first second, not into ten, but into 100 parts. We should then have had to work out the sum of 100 numbers, beginning with 0, 0·0001, 0·0002 ... and ending with 0·0098, 0·0099. Can we shorten the work, so that we shall not actually have to carry out the addition? Yes, this is possible. The first number, 0, and the last number, 0·0099, add up t: 0·0099. The second number, 0·0001, and the next to last, 0·0 98, also add up to the same amount. Going on in this way we can arrange all the numbers in pairs, each pair adding up to 0·099. There will be 50 such pairs. The total will be 50 times 0·009·, that is, 0·495. This result was quoted, without proof, in Chapter 12.

Geometrical Progressions

A geometrical progression consists of a series of numbers, each of which is obtained by multiplying the previous one by a fixed number – e.g., 1, 2, 4, 8, 16 ... or 3, $1\frac{1}{2}$, $\frac{3}{4}$, $\frac{3}{8}$ ... Such series can arise in a number of ways.

For instance, there is the well-known question: at what time between 3 o'clock and 4 o'clock is the minute hand of a clock over the hour hand? It is quite natural to begin thinking in the following way. At 3 o'clock the minute hand is 15 minutes behind the hour hand: the hour hand moves slowly, so that in 15 minutes' time the minute hand will nearly have caught up with the hour hand. The hour hand moves through 5 minutes in each hour – one-twelfth as fast as the minute hand. By 3.15 the hour hand will have moved $\frac{15}{12}$ minutes – this is the amount the minute hand still has to catch up. The minute hand will reach this position after an extra $\frac{15}{12}$ minutes. But meanwhile the hour hand will have moved through another $\frac{15}{12^2}$ minutes. In this way we keep revising our original guess of 15 minutes, by adding to it in turn, $\frac{15}{12}$, then $\frac{15}{12^2}$, and so on – each correction being one-twelfth the size of the previous one. In this way we obtain the result $15+\frac{15}{12}+\frac{15}{12^2}+\frac{15}{12^3}+$ etc., the sum of a geometrical progression. By taking sufficiently many terms of this series, we can get an answer correct to any degree of accuracy: for instance, the four terms actually written above give an answer that differs from the correct answer by less than $0\cdot001$.

It is possible to see what the sum of the series is. The hour hand goes only 5 minutes while the minute hand goes 60 – that is, the minute hand gains on the hour hand at the rate of 55 minutes in each hour, or $\frac{55}{60}$ minute in every minute. $\frac{55}{60}$ is the same thing as $\frac{11}{12}$. Accordingly, to catch up 15 minutes on the hour hand, the minute hand will require 15 divided by $\frac{11}{12}$ minutes – that is, $16\frac{4}{11}$ minutes. So the sum of the series must be $16\frac{4}{11}$.

Another problem: if a ton of seed potatoes will produce a crop of 3 tons, which can either be consumed or used again as seed, how much must a gardener buy, if his family want to consume a ton of potatoes every year for ever?

First of all, he must buy a ton to meet his needs for this year. To get a ton for next year, it will be sufficient to plant $\frac{1}{3}$ ton now. To meet the needs of the year after next, $\frac{1}{9}$ ton will be sufficient: for it will yield $\frac{1}{3}$ ton next year, and this planted again will yield 1 ton the year after. And so on. To meet the needs of his family for ever, the farmer must plant $1+\frac{1}{3}+\frac{1}{9}+\frac{1}{27}+\frac{1}{81}+ \ldots$ tons.

How much does this series add up to? Let us call the number it adds up to x. We can find x by a simple trick – namely, by working out $3x$ and comparing it with x.

Thus $\qquad x=1+\frac{1}{3}+\frac{1}{9}+\frac{1}{27}+ \ldots$

so $\qquad 3x=3+1+\frac{1}{3}+\frac{1}{9}+ \ldots$

We notice that $3x$ is the same series as x, except that 3 has been added at the front, so $3x=3+x$. It follows that $2x=3$, so x must be $1\frac{1}{2}$.

We can easily see that this is the right answer. If the gardener buys $1\frac{1}{2}$ tons, he needs 1 ton to eat this year, and has $\frac{1}{2}$ ton to plant. The crop will be three times as much as what is planted – this is, it will be $1\frac{1}{2}$ tons – and again he has 1 ton to eat and $\frac{1}{2}$ ton to plant. In this way he and his descendants can continue as long as they care to do.

The same type of series arises in connexion with annuities, compound interest, discount, stocks and shares, etc. Compound interest is one of the main historical reasons why geometrical progressions first came to be studied. To a man making a fortune by money-lending, it is, no doubt, an absorbing subject: for most other people, and particularly for children at school, compound interest is likely to prove deadly dull.

Another application of geometrical series is for the study of air resistance. A body moving through the air is like a man rushing through a crowd. The faster he runs, the more people he knocks into: in other words, the resistance to his progress is proportional to his speed. The same is true for a body moving through the air (provided its speed is not too great): the faster it goes, the more air it has to brush out of its way each second. The result is that it loses a definite fraction of its speed each second. If two-thirds of the speed is lost each second, one-third remains: thus a body might move 1 foot in the first second, $\frac{1}{3}$ foot in the following second, $\frac{1}{9}$ foot in the third second, and so on. The distance gone altogether would be $1+\frac{1}{3}+\frac{1}{9}+\frac{1}{27}+\ldots$, the same series as we had before. It is, of course, taken for granted that no force is acting on the body, apart from the resistance of the air. Thus, the body might be a propeller; if it is properly balanced, and is not connected to an engine, there is no force to make it turn round; if it is given a push, it will start spinning, but will gradually slow down, in the way described.

The connexion between a moving body and geometrical progressions was known already in the seventeenth century. A more modern application of the same idea is to an electric current in a wire: an electron moving inside the wire collides with the atoms composing the wire, just like a man moving in a crowd. If the wire is connected to an electric battery, the problem is altered: there is then a force dragging the electron forward. In the same way, a falling raindrop is subject to the pull of gravitation; for this reason, it is not brought to rest by the resistance of the air. The problem of the way in which a raindrop falls is therefore slightly more complicated, but it too was solved, by means of geometrical progressions, in the seventeenth century.

If x stands for 'any number', we can show (by the method used in the potato problem) that the series $1+x+x^2+x^3+\ldots$ is equal to $\dfrac{1}{1-x}$, provided x is not bigger than 1.

Other Series

We have just seen that $\dfrac{1}{1-x}$ can be expressed in the form of a series, containing the various powers of x. This is not a peculiarity of $\dfrac{1}{1-x}$: almost any function of x you are likely to meet can be expressed in this way. For instance, $\sqrt{1+x}$ is equal to a series which begins with $1+\frac{1}{2}x-\frac{1}{8}x^2\ldots$ and $-\log_e(1-x)$ is equal to the series $x+\frac{1}{2}x^2+\frac{1}{3}x^3+\frac{1}{4}x^4+\ldots$ For both these series we suppose x to be less than 1. It is obvious that the series are not true when x is bigger than 1. Of course, we have not *proved* the series to be equal to the functions stated: for proofs you will need to consult text-books.

It is often very convenient to express a function in the form of a series. For instance, while we know from Chapter 12 what is meant by $\log_e 2$, it may not be easy to say just what number this is. By means of the series we can find it. For $\log_e 2$ can be found from $\log_e \frac{1}{2}$. In fact, 2 times $\frac{1}{2}$ equals 1. Taking logarithms, it follows that $\log_e 2 + \log_e \frac{1}{2} = \log_e 1$. But $\log_e 1 = 0$. So $\log_e 2 = -\log_e \frac{1}{2}$. But putting $x = \frac{1}{2}$ in the series given, it follows that $-\log_e \frac{1}{2}$ is equal to $\frac{1}{2} + \frac{1}{8} + \frac{1}{24} + \frac{1}{64} + \ldots$ The terms of this series get small quite rapidly: we do not have to take very many terms to get quite a good value for $\log_e 2$.

Another advantage of such series is that they are easy to differentiate or integrate, since we know how to deal with powers of x. If you differentiate the series for $-\log_e (1-x)$, what series do you get? Is this result reasonable?

Later in this chapter we shall find a series for e^x, and we are now going to find series for $cos\ x$ and $sin\ x$, in order to show how such a question can be tackled.

In Chapter 13 we showed $sin\ 0$ to be 0, and $cos\ 0$ to be 1, also that $\dfrac{d\ sin\ x}{dx} = cos\ x$ and $\dfrac{d\ cos\ x}{dx} = -sin\ x$. It is rather surprising that from this information alone we can find the series we want.

If $cos\ x$ is expressed as a series containing the powers of x, certain numbers will occur in the various terms (as the numbers 1, $\frac{1}{2}$, $\frac{1}{3}$, $\frac{1}{4}$, etc., did in the series for $-\log_e (1-x)$): these numbers we shall call for short $a,b,c,f,g,h,j,k \ldots$ (The numbers d and e are left out, because d is used with a special meaning in $\dfrac{dy}{dx}$ and e also has a special meaning.) So the series will be:

$$cos\ x = a + bx + cx^2 + fx^3 + gx^4 + hx^5 + jx^6 + kx^7 + \ldots$$

Our job is to find out what particular numbers a, b, c, etc., are.

a we can find straight away. If we put $x = 0$, $cos\ 0 = 1$, while the series becomes simply a. It follows that $a = 1$.

If we differentiate the equation above, we find (since the differential of $cos\ x$ is $-sin\ x$):

$$-sin\ x = b + 2cx + 3fx^2 + 4gx^3 + 5hx^4 + 6jx^5 + 7kx^6 + \ldots$$

b can now be found by putting $x=0$. $sin\ 0=0$, so it follows that $b=0$.

We now differentiate the series for $-sin\ x$. The differential of $sin\ x$ is $cos\ s$, so we have:

$$-cos\ x=2c+6fx+12gx^2+20hx^3+30jx^4+42kx^5+\ \ldots$$

c is now found by exactly the same method. Put $x=0$. This leads to the equation $-1=2c$, so $c=-\frac{1}{2}$.

It is clear that there is nothing to stop us continuing with this as long as we like, and finding as many of the numbers $f,\ g,\ h,\ \ldots$ as we care to do. The results are (as you can check for yourself):

$$a=1,\ b=0,\ c=-\tfrac{1}{2},\ f=0,\ g=\tfrac{1}{24},\ h=0,\ j=-\tfrac{1}{720},\ k=0:$$

so:
$$cos\ x=1-\tfrac{1}{2}x^2+\tfrac{1}{24}x^4-\tfrac{1}{720}x^6\ \ldots$$

The rule which gives the numbers 1, 2, 24, 720, etc., which appear here, is the following. We start with 1. We multiply this by 1 times 2; this gives the second number, 2. We multiply the second number by 3 times 4, this gives the third number, which again is multiplied by 5 times 6 to give the fourth number: and so on. Differentiating the above series, we find the series for $sin\ x$:

$$sin\ x=x-\tfrac{1}{6}x^3+\tfrac{1}{120}x^5-\ \ldots$$

These are good series for the purpose of calculation, since the terms get smaller very rapidly, and the first few terms of the series give quite accurate answers. These series therefore give an answer to the problem put in Chapter 13, to find a way of making a table of sines and cosines without drawing any figures.

The Dangers of Series

Series played an important part in the early days of the calculus, particularly in the years following 1660. This was a period of great practical activity: men were interested in the new developments of science, and were faced with a great variety of practical problems – the construction of clocks and of telescopes, of maps and ships. If a mathematical method gave the correct answer to a practical problem, people did not bother much whether it was logical or not. In dealing with small changes, Δx, mathematicians followed their own convenience: at one moment they said, 'Δx

is very small, it will be convenient to regard Δx as being equal to 0.' A little later they wanted to divide by Δx, so they said, 'If Δx is 0 we cannot divide by it: we will suppose Δx to be small, but not quite 0.' Whichever was more convenient, that they supposed to be true. If the answer turned out to be wrong, they scrapped their work. As the results were always compared with practice, this rough-and-ready method worked quite well.

Series were treated in this way, too. If it looked reasonable to make a certain step, that step was made. If it gave a ridiculous answer, one soon recognized that something was wrong.

After about 150 years of carefree mathematics, difficulties began to be felt. For instance, in the calculation of logarithms, the series $1 - \frac{1}{2} + \frac{1}{3} - \frac{1}{4} + \frac{1}{5} - \frac{1}{6}$ ... arises. Half the terms of this series have a $+$ sign; we will call the sum of these terms a, so that $a = 1 + \frac{1}{3} + \frac{1}{5} + \frac{1}{7}$ ... We will call b the sum of the other numbers in the series – that is, $b = \frac{1}{2} + \frac{1}{4} + \frac{1}{6} + \frac{1}{8} +$ ... We notice that every number that occurs in b is even. If we double b we thus have $2b = 1 + \frac{1}{2} + \frac{1}{3} + \frac{1}{4} +$ ...; in the series for $2b$ we have all the terms of the series for a, together with those for b. So it seems that the series for $2b$ should equal $a + b$, so that $2b = a + b$. It follows from this that $b = a$. But b cannot equal a, for every term in a is bigger than the corresponding term in b; 1 is bigger than $\frac{1}{2}$, $\frac{1}{3}$ is bigger than $\frac{1}{4}$, $\frac{1}{5}$ is bigger than $\frac{1}{6}$, and so on. If a equals b, our original series $1 - \frac{1}{2} + \frac{1}{3} - \frac{1}{4}$ ... $= a - b = 0$. But the sum of the original series is, in fact, just less than $0 \cdot 7$.

In other words, by doing things that *look* reasonable, we have been led to an untrue result. On the other hand, in many cases true and useful results have been obtained by the use of series. It is therefore natural that mathematicians should have begun to inquire more carefully just what is meant by a series, and just what operations may be carried out with series. During the nineteenth century mathematicians carried out such an inquiry. There was a reaction from the carefree attitude of earlier times, and a spirit of caution spread. Mathematicians became rather like lawyers, very concerned about the exact use of words, very suspicious of arguments which merely 'looked reasonable'. Terms such as 'convergence' and 'uniform convergence' were invented, designed to distinguish series which were reliable from those which led to wrong conclusions.

Mathematicians did not only investigate the logic of series: they became uneasy about all the words they were using, and did not become comfortable again until they had given very exact explanations of all the terms they used. Modern books on mathematics are often much longer than the old books, because they spend time explaining and justifying things, which at first sight seem obvious.

There is a fable about a centipede which was asked in what order it moved its feet, and became so puzzled by the question that it was unable to walk at all. Students who begin by studying modern mathematics often suffer from a similar disorder: they spend so much time learning how to criticize, that they never understand how to create. The best policy is to follow the course of history: first to learn to *see* results as the old pioneers were able to see them, only later to examine the weaker points in the natural approach. If there had been no risks taken by the creative mathematicians in the seventeenth and eighteenth centuries, there would have been nothing for the pure mathematicians to criticize in the nineteenth.

The Background of e^x.

Many text-books give explanations of the number e which in themselves are excellent and logical, but leave the reader with the feeling that everything has come 'out of the blue': the argument is logical, but how was it discovered? What is it all about?

We have already had to deal with expressions such as 10^x, a^x, $\log_a x$. We shall now try to collect the facts about these expressions, and to show the connexions between them.

The idea of an exponential function springs immediately from the practice of money-lending. The way in which debt mounts and strangles its victim is an old story, both in fact and fiction. If a money-lender advances £100 now in return for £110 in a month's time, and in a month's time the borrower cannot pay, he finds himself obliged to contract a new loan on the same terms for another month – but the new loan is for £110, not for £100. In a year the debt will become something over £313. The debt for each successive month contains an extra multiplication by $1\frac{1}{10}$. (Compare this with the section 'How Logarithms were invented' in Chapter 6.) 10% a month is the same as 213% a year, much more than 12 times 10%.

We could reverse this and ask: what rate per month is equal to 5% per year? We could try different rates per month, until we found one which worked out (to a sufficient degree of accuracy) to 5% a year. We could ask what rate a week, what rate a day corresponds to 5% a year. If we liked, we could find the rate per hour, or per minute or per second. There would be only one correct answer to any of these questions. By fixing the rate of interest for a year, we automatically fix the rate of interest for any other length of time.

Suppose, for instance, the rate for a whole year to be 100%, and that an inexperienced money-lender charged 40% for six months. Then nobody would borrow money by the year. It would be cheaper to borrow for two periods of six months. £100 now would mean paying £140 after six months. This debt could be met by starting a new loan, for £140. The interest for the remaining six months, at 40%, would be £56, so that £196 would have to be paid at the end of the full year. Borrowing for a year at a time, £200 would have to be paid. In the same way, if the rate for six months were fixed at 50%, it would pay people to borrow money for a year, and lend it out again for two periods of six months: in the first six months, £100 would become £150: in the second six months, £150 would become £225; after paying back £200, a profit of £25 would be left. From practical necessity, then, the rate for six months would be something more than 40%, but less than 50%.

We could make a table showing what £1 would become after any length of time – weeks, days, hours, minutes – once the rate of interest for a year was given. If £1 becomes £a in one year, in n years (n standing for any whole number) it will become £a^n. It is therefore natural to suppose that £$a^{\frac{1}{2}}$ represents the amount that £1 would become after $\frac{1}{2}$ year. A sign such as $a^{\frac{1}{2}}$ has no meaning in itself: in the South of England 'stack' usually means a haystack, in the North a chimney – it is a waste of time to argue which is 'right'. A word is a label tied to a real thing for convenience. It does not matter whether the label is pink or green. A rose by any other name would smell as sweet. If we like to say that £$a^{\frac{1}{2}}$ is going to mean what £1 becomes in $\frac{1}{2}$ year under certain conditions, we have a perfect right to do so. (This definition agrees with that given in Chapter 6, although it uses a different picture.) £a^x will mean what £1 becomes after any number, x, years: x may be a fraction.

As we saw in Chapter 6, the most convenient way of making a table is to start with a small change, and build up from this to larger changes. At whatever rate of interest £1 grows, there will be a time in which it increases by one-thousandth, say, in k years (k may be a small fraction). Every k years that pass, the amount due will be multiplied once again by $1\frac{1}{1000}$. We can thus draw a graph, showing the growth of the debt, by taking upright lines, each separated from the next by a distance k inches. Each line must be longer than the previous line by one part in a thousand. Corresponding to $x=0$ we must have a height of 1 inch: since the sum of money, to begin with, is £1.

Minus Numbers

It is clear that we could extend our graph to minus values of x. Each line is $\frac{1000}{1001}$ of the line to the right of it. So, k inches to the left of $x=0$, we could put a line having the height $\frac{1000}{1001}$, and continuing in this way, we could continue the graph, and find a height corresponding to any distance to the left – that is, to minus values of x. We now have a graph extending as far as we like, both to the right and to the left.

Change of Scale

The distance k depends on the rate of interest. By a suitable choice of k we can get any rate we wish. The distances between our upright lines must remain *equal*, but by altering their size we change the rate of interest. Mechanically this can be done by the arrangement known as 'lazy tongs'. In Fig. 23 we indicate this by a lattice-work of diamond shapes. We suppose a model made to this design, with loosely jointed pieces of wood. Some device (not shown in the figure) will be needed to keep the upright rods in the proper direction. By pulling or pushing at the points A and B, the upright rods can be spaced out, or brought more closely together. In this way, we have one model which represents a^x for *any* number a (within a certain range of values). (In the

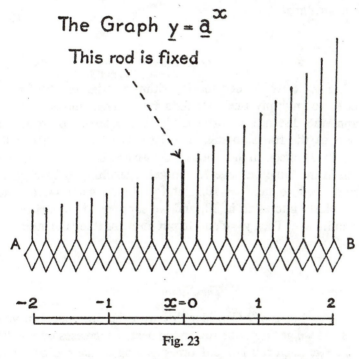

Fig. 23

figure, the rate of change has been exaggerated – each upright stick is actually one-tenth part greater than its neighbour, instead of one-thousandth.) x is throughout measured in inches. When $x=1$, $a^x=a$. Thus a is the length of the rod standing at a distance 1 inch to the right of $x=0$. For example, to obtain the graph of 2^x we must press the points A and B until the rod 2 inches long stands above the point marked 1 on the scale for x. When the model is set in this way, we shall find that 1 inch to the right of any rod there stands a rod twice as high.

The model actually drawn – that is, one in which each rod is $1\frac{1}{10}$ as long as its neighbour on the left – we shall call the *crude* model: the one described in the text, using the ratio $1\frac{1}{1000}$, we shall call the *fine* model. The crude model is suitable for actual construction, and for class work: the fine model should stay in the imagination, for the purpose of argument. As we saw in Chapter 6, the ratio $1 \cdot 1$ is not sufficiently near to 1 to give accurate values for logarithms.

Logarithms

In Chapter 6 we defined the logarithm as 'the length of rope' needed to multiply one's strength by a given number. In our graph, the distance x corresponds to the length of rope, the height of the rod standing there (y inches, say) measures the multiplying effect. In the crude model we use, in fact, the numbers given in the table on page 75. To get logarithms to base 10, we would need to take $a=10$. But for the moment we are not particularly interested in 10, and we suppose the model set for any number a. Then $y=a^x$, or, to put the same thing the other way round, $x=\log_a y$.

The Number e

If $y=a^x$, what is y'? Consider this with the fine model in your mind. As we go from one rod to the next, x increases by k – that is, $\Delta x=k$. Each rod is $1 \cdot 001$ times as long as the rod before it; the change in length, Δy, will therefore be $0 \cdot 001$ times y. So $\dfrac{\Delta y}{\Delta x}$ will be $\dfrac{0 \cdot 001}{k} y$. This gives us some idea of y': it suggests (what is actually true) that y' is proportional to y. If we take $k=0 \cdot 001$, we shall have a particularly simple result. $\dfrac{0 \cdot 001}{k} y$ will then be simply y, and we shall have, very nearly, $y'=y$. (We have to write 'very nearly', because $\dfrac{\Delta y}{\Delta x}$ is 'very nearly', but not quite, y' when $\Delta x = 0 \cdot 001$.)

Taking $k = 0 \cdot 001$ means that the rods are spaced with one-thousandth of an inch between them. In going from $x=0$ to $x=1$, the length of the upright rod will have been multiplied by $1 \cdot 001$ a thousand times. So a, the length of the rod at $x=1$, will be $(1 \cdot 001)^{1000}$.

If instead of the fine model we had argued from the crude model, we should have been led to the result $(1\frac{1}{10})^{10}$. The fine model gives the better answer, $(1\frac{1}{1000})^{1000}$. By taking still larger numbers of rods we could get still better answers. Our result would always be of the form $\left(1+\dfrac{1}{n}\right)^{n}$. The larger n is, the more nearly would y' be equal to y. As n is made larger and larger, $\left(1+\dfrac{1}{n}\right)^{n}$ gets closer and closer to the number $2 \cdot 71828 \ldots$, which was mentioned in Chapter 11, and is named e. If $y=e^x$, $y'=y$ exactly.

In Chapter 11, e was found by another method – namely, by choosing a number a which gave the simplest result for the differentiation of the logarithm to base a. As $y=a^x$ means the same thing as $x=\log_a y$, it is not surprising that the same number e should give the simplest answer in both cases. The reader may be able to show that the two methods are really the same. The only complication is due to the fact that the signs x and y have changed places: in Chapter 11, we assumed $y=\log_a x$: here, $x=\log_a y$.

The Series for e^x

We now have enough information about e^x to find a series for it. When $x=0$, $e^x=1$. If $y=e^x$, $y'=e^x$. The method that was used to find a series for $cos\ x$ works equally well for e^x. We find, in fact:

$$e^z = 1 + x + \tfrac{1}{2}x^2 + \tfrac{1}{6}x^3 + \tfrac{1}{24}x^4 + \tfrac{1}{120}x^5 + \tfrac{1}{720}x^6 + \cdots$$

Differentiate this series for yourself, and verify that the series for y' is the same as for y, so that $y' = y$.

This method works also for many other functions, and is associated with the names of Taylor and Maclaurin.

a^z has simple properties, similar to those of e^z, for, as we have seen, the graph of a^z can be got from that of e^z simply by changing the scale of x (in the model, by pushing or pulling the points A and B).

The Applications of e^z

The importance of e^z is due to the property $y' = y$: that is to say, the rate at which it grows is equal to its size. If instead of e^z we take e^{bx}, where b is any fixed number, we have $y' = by$; that is, the rate of growth is proportional to the size.

There are many things which grow in this way. We have already mentioned the example of money-lending. £1,000 grows a thousand times as fast as £1.

Much the same thing holds in business. Within certain limits, the more shops a company owns, the more rapidly can it extend its business.

If a country desires to build up its industry, and starts with very little equipment, it finds that the rate at which it can install new factories is very slow; but the more factories it gets, the quicker it can equip new factories. In a country suffering foreign conquest, the reverse holds: the more factories it loses, the less it is able to replace its losses.

The population of a country, under settled conditions, may grow according to the law e^{bx}. The more people there are in the country, the more children are likely to be born. The population of the U.S.A. between 1790 and 1890 roughly corresponded to the formula $y = 3 \cdot 9 \times 10^{0 \cdot 012x}$, where y is the population in millions x years after 1790.* The formula, of course, ceases to work when a

* *Introduction to Mathematics* by Cooley, Gans, Kline, and Wahlert, page 363.

country reaches the stage where it cannot support any more people. Rather similar considerations apply to the rate at which microbes multiply in a glass of milk that is going sour, the spread of rabbits in Australia, and other forms of living growth.

There are also conditions in which a new religion or political creed grows by an exponential law. If there are large numbers of people in the mood to accept some new doctrine, once it is put to them, the spread of that doctrine will largely depend on the number of men and women who act as missionaries for it. So long as Mahomet is a lonely man, he can only influence those in his own district. With every convert his power to make himself heard increases. It is possible to find cases where statistics show that a movement has grown according to an exponential law, subject of course to slight variations, due to other causes and particular events which helped or hindered the movement. The fact that a movement grows in this way *during a certain period* tells us nothing at all about its future prospects: it may be smashed by bad leadership, or by disillusion, or by superior force, or by sheer bad luck. When human events show a mathematical law, it means that one or two simple causes were decisive at a certain time: the more different causes are at work, the more complicated will the graph of the movement become.

But the place in which an exponential function really feels at home is far from the complications of human or animal life. In the sciences of non-living matter, exponential functions abound; the speed of a body moving against air-resistance, the pressure of the atmosphere at different heights, the vibrations of an electric circuit, the passage of electrons through a gas, the decay of radium, the speed of a chemical reaction, the current in an electro-magnet, the dying away of any vibration – in these and in countless other problems some quantity grows or shrinks at a rate proportional to its size. It is indeed remarkable how much of the physical world, amid the conflicting action of a great variety of unconnected forces, can be described by the simplest mathematical functions, x^n and e^x.

CHAPTER 15

THE SQUARE ROOT OF MINUS ONE

'*The prevalent idea of mathematical works is that you must understand the reason why first, before you proceed to practise. That is fudge and fiddlesticks. I know mathematical processes that I have used with success for a very long time, of which neither I nor anyone else understands the scholastic logic. I have grown into them, and so understand them that way.*' – Oliver Heaviside.

At the end of Chapter 5 we saw that the square of every number had a $+$ sign, so that no number could exist with a square equal to -1. One would naturally expect the matter to rest here, and mathematicians to admit that any problem which led to the equation $x^2 = -1$ was meaningless and had no solution.

But a strange thing happened. From time to time mathematicians noticed that their work could be much shortened, and the correct answer obtained, if in the middle of the working they used a sign i, assumed i^2 to be -1, and in all other respects treated i just as if it were an ordinary number. This was first done about 1572. The people who did it were very doubtful about the method, but it kept on giving the correct answers. Nobody knew why it should, but the new sign i proved so useful that for two centuries mathematicians used it, without any justification other than success. It was not until 1800 that a logical explanation of the meaning of i was given. (The whole story will be found in Dantzig's *Number, the Language of Science.*)

If, for the moment, we allow i to be treated as an ordinary number, we can see the type of result obtained by eighteenth-century mathematicians.

In Chapter 14 we found series for $e^x \cos x$, and $\sin x$. You may have noticed that the same numbers came into these series. In fact, if you take the series for e^x:

$$1+x+\tfrac{1}{2}x^2+\tfrac{1}{6}x^3+\tfrac{1}{24}x^4+\tfrac{1}{120}x^5+\tfrac{1}{720}x^6 \ldots$$

and leave out every other term:

$$1+\tfrac{1}{2}x^2+\tfrac{1}{24}x^4+\tfrac{1}{720}x^6+ \ldots$$

and then make the signs alternately + and —

$$1-\tfrac{1}{2}x^2+\tfrac{1}{24}x^4-\tfrac{1}{720}x^6 \ldots$$

you obtain the series for *cos x*. In the same way, *sin x* corresponds to the other half of the terms.

By making use of the sign *i* we can express the relation between the three series in a short formula.

Let us suppose *x* to have the value *ia*, where *a* may be any number. Putting $x=ia$ in the series for e^x we have:

$$e^{ia}=1+ia+\tfrac{1}{2}i^2a^2+\tfrac{1}{6}i^3a^3+\tfrac{1}{24}i^4a^4+ \ldots$$

i^2 we know is —1, $i^3=i\times i^2=-i$. $i^4=i^2\times i^2=(-1)\times(-1)=+1$. In the same way, all the higher powers of *i* turn out to be 1, *i*, —1, and —*i* in turn. Accordingly,

$$e^{ia}=1+ia-\tfrac{1}{2}a^2-\tfrac{1}{6}ia^3+\tfrac{1}{24}a^4+ \ldots$$

If we sort out the terms which contain *i* from those which do not, we see that the terms without *i* give the series for *cos a*, while the terms containing *i* are equal to *i* times the series for *sin a*. In short:

$$e^{ia}=cos\ a+i\ sin\ a$$

This is rather a surprising result. e^x is a simple type of function, like 2^x or 3^x. We meet functions of this type very early – in the arithmetic course, for compound interest. *e* is just a number between 2 and 3, round about 2·7.

cos a and *sin a* are quite different. We meet them first of all in connexion with geometry, as the sides of a right-angled triangle. We have no reason at all to expect that they will be connected with any of the simpler formulae of algebra: in fact, it is a mystery to most people how the tables for *sin a* are worked out at all.

The formula above shows that *cos a* and *sin a* are in fact closely connected with the simplest type of function. e^x has many simple properties; for instance, $e^p\times e^q=e^{p+q}$, *p* and *q* standing for any two numbers. If we take $p=ia$ and $q=ib$ we obtain the results below:

$$e^{ia}\ e^{ib}=e^{i(a+b)}$$

that is $(cos\ a+i\ sin\ a)\ (cos\ b+i\ sin\ b)=cos\ (a+b)+i\ sin\ (a+b)$. If we multiply out the expression on the left-hand side, and compare the two sides, we see that $cos\ (a+b)$, the part free from i, must be the same as $(cos\ a\ cos\ b—sin\ a\ sin\ b)$, while the number that appears with i, $sin\ (a+b)$, must be the same as the number that appears with i on the other side, $(cos\ a\ sin\ b+sin\ a\ cos\ b)$.

All the formulae given for sines and cosines in books on trigonometry can be obtained, usually without very much work, from the properties of e^z. This fact can be used to take a burden off the memory. Instead of learning the formulae, one can work them out whenever they are needed, by using e^{ia}.

It is easy to find formulae giving $cos\ a$ and $sin\ a$ in terms of e^{ia}. $cos\ a$, in fact, equals $\frac{1}{2}(e^{ia}+e^{-ia})$. We can turn any problem about cosines into a problem about exponentials. For instance, we can immediately find $\int (cos\ x)^6 dx$ by this means, for exponential functions are easy to integrate.

We can regard all problems about sines and cosines as problems about exponentials, thus saving ourselves the trouble of learning special methods for dealing with sines and cosines. i is therefore a very helpful device: as was mentioned in Chapter 5, electrical engineers make great use of it.

What is i?

It seems at first sight very strange that the square root of minus one – something which no one has ever seen, and which seems in its own nature to be impossible – should be useful for such material tasks as the design of dynamos, electric motors, electric lighting, and wireless apparatus.

When some natural fact strikes us as strange, it means that we are looking at it from the wrong point of view. If we find the universe mysterious, it is because we have some idea about what the universe ought to be like, and are then surprised to find it is something different. The fault lies with our original idea – not with the universe.

When we find i mysterious, it is because we are thinking of i as an ordinary number. But there is no *number* x for which $x^2=-1$. We convinced ourselves of this in Chapter 5.

We have also seen that the sign i, for which $i^2=-1$ is assumed, leads to perfectly correct results, and clearly has *some* meaning. It is impossible that i should be a *number*; but there is no contradiction at all if we suppose i to be something else. i, in fact, can be interpreted as an operator.

An operation means doing something: 'Turn the piano upside down', 'Move two paces to the right', 'Throw Mr Jones out' are examples of operations applied to a piano, a soldier, and Mr Jones. If we use **U** as an abbreviation for 'turn upside down', and p for 'the piano', **U**p has the same meaning as the first sentence given above. **U** is called an *operator*. The operation can be repeated. **UU**p would mean 'Turn the piano upside down, and then turn it upside down again', which would, of course, bring the piano back to its original position. Usually **UU** is represented by **U²**, **UUU** by **U³**, etc. Since turning the piano upside down twice leaves it in its original position, **UU**$p=p$, or **U²**$p=p$. It is convenient to use the sign **1** for the operation of leaving something alone. Thus **1**p means the result of leaving the piano alone, which is just the piano in its original position, p. So **U²**$p=$**1**p. This sort of result is not only true for a piano; the result of turning any solid body (not a glass of water, however!) upside down twice is the same as that of leaving it alone. We express this by the equation **U²**=**1**.

You will see that it is perfectly possible to argue about operations, and to get results about them, the truth of which can be seen purely by common sense. You will also see that these results, when stated in the shorthand of algebra, *look* like equations for numbers, and might easily be mistaken for statements about numbers. And in fact it is precisely this mistake which has been made in connexion with the equation $i^2=-1$. To avoid any such misunderstanding, we shall print all signs representing operations in heavy type. From now on, in particular, we shall write i instead of i, so as to make it clear that we are *not* dealing with the sign for a number.

While operators are not numbers, they are often closely

connected with numbers. In an adding machine, for instance, we have a number of gear wheels arranged in the same way as the wheels in the mileage recorder of a motor-car. Every time a motor-car goes a mile, the wheel representing units turns through one division, and one is added to the mileage. The turning of the wheels is an operation, and this operation corresponds to adding one to the mileage. It is because of this correspondence between numbers and certain mechanical operations that it is possible to make calculating machines at all.

We shall now find a set of operators **1, 2, 3** ... and **+** which correspond very closely indeed to 1, 2, 3 ... and + in ordinary arithmetic. **1, 2, 3** ... are not numbers, but there are many relations between these operators which correspond to those between ordinary numbers: they have a pattern in common with ordinary numbers, just as a family of father, mother, son, and daughter in a colony of chimpanzees have a pattern in common with a family of father, mother, son, and daughter in Birmingham. This is not to say that everyone in Birmingham is a chimpanzee.

We shall then find it quite natural to bring in an operation, **i**, such that $i^2 = -1$.

The Operators **1, 2, 3** ....

To define the operators **1, 2, 3** ... we begin by imagining a long lath of wood, on which a scale has been marked. O is any fixed point and the figures 1, 2, 3, etc., are marked at distances of 1, 2, and 3 inches to the right. Going to the left from O, we find the figures —1, —2, —3, etc., at distances 1, 2, 3 inches. This is an ordinary scale, such as might be marked on a thermometer.

One end of a wire is fastened to O, and on this slides a bead, A. The wire may point either to the right or the left. The operations we shall consider will consist either in turning the wire from one direction to the other, or in sliding the bead A along the wire.

The operation **2** may now be defined. It consists in moving the bead A to a point on the wire twice as far from O as the former position of A. The operation **2** may be put into words, 'Double the distance OA'. In the same way, the operation **3** means

'Make OA 3 times as long.' $2\frac{7}{8}$ means, 'Make OA $2\frac{7}{8}$ times as long.' **x** means 'Make OA x times as long', where x stands for any positive number. **1** will mean, 'Leave A where it is.'

Several operations may be carried out one after the other. For instance, the operation **(4) (3) (2)** means that the length OA has to be doubled, then trebled, then again made four times as large: in short, OA has to be made 24 times its original length. The three operations, applied in turn, are equivalent to the single operation **24. (4) (3) (2)**=**24**. So there is a close correspondence between the doing in turn of various operations and the multiplication of ordinary numbers. We might say that the operations have the same multiplication table as ordinary numbers.

By the operation **—1** we understand that the direction of the wire is reversed, but the distance OA is left unchanged. Thus, if A was originally above the mark 3, the operation **—1** would cause it to come over the mark —3. If A was originally at —3, the operation **—1** would bring it to 3.

By **—x** we understand that the distance OA is made x times as large, and its direction reversed.

Check for yourself that **(—2) (3)**=**—6** and **(—4) (—5)**=**20**. The rules for multiplying **+** and **—** operators are the same as those for ordinary numbers.

Addition

How shall we find a meaning for **2+3** or **2+ (—3)**? We *might* say straight away that **2+3** is to be **5** and that **2+ (—3)** is to be **—1**; that is, we could use ordinary arithmetic as a way of defining addition for operators. But this would have the disadvantage that when we come to consider **i**, which does not correspond to any ordinary number, we should not know what to take for **1+i**.

It is therefore more satisfactory to make use of another method which will apply equally well to operations such as **i**, while agreeing with the first method for operators that correspond to ordinary numbers.

We suppose the bead A to be at any point P, to begin with. Let

Q be the point to which the operation **2** sends A, R the point to which the operation **3** sends A, and S the point to which A is sent by the operation **5**. Then $OQ=2. \ OP$; $OR=3. \ OP$; $OS=5. \ OP$. (OP stands for the distance from O to P, 2, 3, 5 for the ordinary numbers; there are no operators in these equations.) it is obvious that $OS=OQ+OR$, so that we could find the position of S by putting lengths equal to OQ and OR end to end.

In the same way we could find the effect of the operation **2+(—3)**. We must remember that **2** and **—3** will cause OA to point in opposite directions: when we put the two lengths end to end, it must be in such a way that the second length points in the opposite direction to the first. Fig. 24 may help to make this clear.

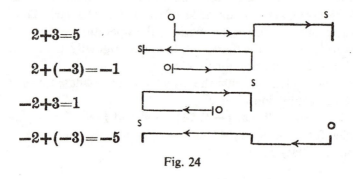

2+3=5

2+(—3)=—1

—2+3=1

—2+(—3)=—5

Fig. 24

Accordingly we are led to define addition as follows: if the operation **x** sends A from P to Q, and the operation **y** sends A from P to R, **x+y** is defined as the operation which sends A to S, where S is the point obtained by putting OQ and OR end to end.

We may write **2+ (—3)** more shortly as **2—3**. Be careful to distinguish between **2—3** and **(2) (—3)**. **(2) (—3)** means that the operation **2** has to be applied to the result of **—3** acting on A.

We have now found a set of operators which correspond completely to the ordinary numbers: they can be multiplied and added, and the answers look exactly like the answers for ordinary numbers, except that they are in heavy type. If you picked up a page of calculations, dealing with these operators, you might mistake it for examples in elementary arithmetic, written by someone who pressed very hard on the pen: there would be no way of distinguishing between the two.

The Operator i

The operation −1 has the effect of reversing the direction of OA, without altering its length – i.e., it rotates OA through 180°.

Can we find an operation i such that $i^2 = -1$? i^2 means that the operation i is carried out twice. The question asks us to find an operation which, performed twice, turns OA through 180°.

The question is now ridiculously simple. The mysterious operation i consists simply in turning OA through 90°! In Fig. 25 A is supposed to start by being at P. The operation i sends A to E, i^2 sends A to F, i^3 sends A to G, and i^4 brings A back again to P. −1 sends A from P to F, and $i^2 = -1$, as we hoped.

Before we brought i into the picture, the bead A moved on a straight line. It could lie to the right or the left of O, but always

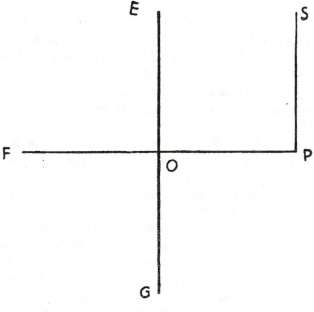

Fig. 25

at the same level as O. Now that i has been brought in, it is possible for A to lie above or below O, and in fact we shall soon have A wandering over the whole surface of the paper.

Addition

The 'end-to-end' method of addition can now be used to give a meaning to expressions such as **1+i** and **2+3i**.

Suppose, as before, that the bead A is at P, and is then acted upon by the operation **1+i** (Fig. 25). Where will **1+i** send it? We have to find the points Q and R to which **1** and i send A, and then put OQ and OR end to end. The operation **1** leaves A at P. The operation i sends A to E. So Q is P and R is E. We have to put OP and OE end to end. At P we draw PS equal to OE, and having the same direction as OE. This gives us the point S, which we require. The operation **1+i** sends A from P to S. If A starts at any other point, we can find where the operation **1+i** sends it. (For instance, if the bead A were at E to begin with, where would **1+i** send it?)

In the same way, the operation **2+3i** can be studied. The bead A may start at any point of the paper, say at K (Fig. 26). We have to study the operations **2** and **3i** separately, and then to combine them by 'end-to-end' addition.

The operation **2** would send A from K to L. **3i** acting on A means that OA has to turn through 90° and become 3 times as long. Thus **3i** would send A from K to M. Now OM has to be put on to the end of OL. We draw LN equal to OM, and in the same direction as OM. N is the point we are looking for. The operation **2+3i** sends A from K to N.

K may be chosen anywhere. We could choose K in different positions, and notice where the corresponding N came. You may notice that the angle KON is always the same, and that ON is always in the same proportions to OK, wherever K may be chosen, In other words, wherever the bead A may be, the operation **2+3i** turns OA through the same angle and stretches the length OA the same number of times.

We may picture any operation **a+bi** as being a turn through

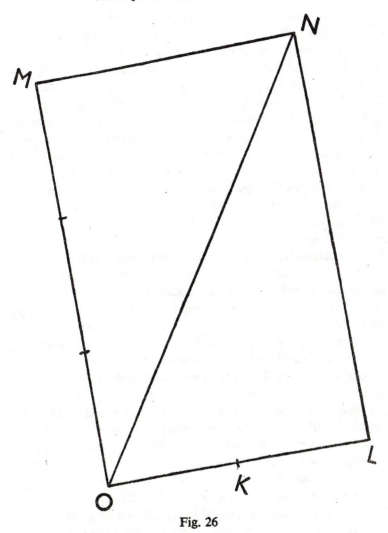

Fig. 26

some angle followed by a stretch. If **a+bi** corresponds to a turn through the angle θ and a stretch r times, it is easy to see that $a = r \cos \theta$ and $b = r \sin \theta$. If we are given a and b, we can find r and θ graphically by drawing a right-angled triangle, with sides a and b. r is known as the *modulus*, θ as the *argument* (sometimes the *amplitude*) of the operation **a+bi**. Special names are given to these two quantities, because they arise naturally and often occur in formulae: we therefore save time by giving them names.

You are now in a position to think about these operations for yourself. We have shown, by the examples **1+i** and **2+3i**, how to find the operation represented by any symbol of the type **a+bi**. You now understand what the symbols represent, and it is up to you to make yourself familiar with the actual operations, by experimenting with them. What do you understand by the operations **1+2i, 1—i, —3i**? If two operations are carried out in turn, does it matter in which order this is done? Is **(2) (i)** the same as **(i) (2)**? Is **(1+i) (1+2i)** the same as **(1+2i) (1+i)**? By carrying out the actual geometrical operations, find a single operation which has the same effect as **(1+2i) (1+i)**. What is the modulus of **i**? Of **3+4i**? What operation is **—i**? Does **(i) (i)=(—i) (—i)**? What is **(i) (—i)**? **—i** means the operation **(—1) (i)**.

Once we have given a definite meaning to the symbols, we lose all control over them. We can decide what name to give any operation, but once we have chosen the name, we have to observe what that operator actually does. We have reached that stage. We have given the names **1, 2, 3 . . .** and **i** to certain operators, and have explained what we mean when two operators are written side by side, or are linked by the signs **+** and **—**. Division is to mean the opposite of multiplication. The only sign which has not yet been given a meaning is e^x, which we shall come to later. But so far as addition, subtraction, multiplication, and division are concerned, everything is settled. We must not *assume* that these new signs obey the same rules as ordinary numbers: for they are not ordinary numbers. We must try to see whether they do.

For example, we must not *assume* that **(2) (i)** is the same as **(i) (2)**. Actually **(2) (i)** is equal to **(i) (2)**, but we must convince ourselves that this is so by trial. There are operators for which the order in which they are performed alters the answer. The effect of being beaten and then beheaded is different from being beheaded and then beaten.

The interesting thing about the operators we are now dealing with is that they behave just like ordinary numbers. If you take any formula which is true for ordinary numbers, it will be true for these operators. For instance, $(x+1)(x-1)=x^2-1$, when x is an ordinary number. If we put any operator **a+bi** in the place of x, we find the result (in heavy type) still holds true. For instance, putting **i** in place of x, it is true that **(i+1) (i−1)= i²−1**. **i²** is **−1**, so **i²−1=−2**. You will find that the result of carrying out the operations **i−1** and **i+1**, one after the other, is to double the length *OA*, and to turn it through 180°; that is, to do what the operation **−2** does.

You will find, too, that it does not matter in what order multiplication is done, or in what order signs are added to each other. **i2** and **2i** have exactly the same meaning (multiplication meaning that the operations are carried out, one after the other) and **i+1** has the same meaning as **1+i** (it does not matter which line is put to the end of the other, in 'end-to-end' addition). In short, any rule which is true for ordinary numbers is true for these operators.

This fact is extremely convenient. Often, when we begin to study a new type of operation, we find laws which are entirely fresh to us. Each type of operation has its own particular way of behaving, which we have to get used to. But we do not have to learn any new rules for the operators **a+bi**. They behave exactly *as if* they were numbers: while they are not, in fact, numbers, they yet have so much in common with numbers that, *for most purposes*, they can be thought of as numbers. Mathematicians usually refer to them as *complex numbers*, to show that they are close relations of the ordinary numbers. If, in making some calculation, you treat **i** as if it were an ordinary number, you will obtain the correct result.

On the other hand, by thinking of **i** as an operator, you can often see results more quickly than by using the methods of ordinary arithmetic. For instance, you may be asked to solve the equation $\mathbf{x}^2=\mathbf{i}$. We know that **i** represents a rotation through a right-angle. We are asked: what operation **x**, carried out twice, has the same effect as **i**? The answer is obvious: turn through half a right-angle. This operation does not involve any stretching, so its modulus $r=1$. (*No stretching* does *not* mean $r=0$. The length *OA* is *multiplied* by r. If *OA* is unchanged, this means $r=1$.) Since the angle θ is half a right-angle, it is easily seen that the numbers a and b must be 0·707 and 0·707 (from a table of sines and cosines), and the operation **a+ib**, which represents a turn through half a right angle, is **0·707+0·707i**. (This is an outline only. Draw the figure for yourself. Also check the result by ordinary arithmetic, treating **i** as if it were an ordinary number.)

Complex Numbers and Electricians

It is now easier to see why electricians make such frequent use of the operator **i**. Every generator of electric current contains parts which are rotating, and which every minute pass through many right angles – that is, have the operation **i** applied to them, again and again.

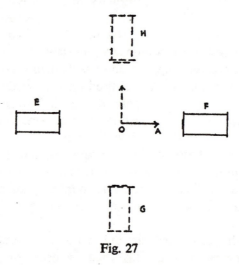

Fig. 27

It would be possible, and for electrical students interesting, to explain **i** entirely in terms of a simple generator of alternating current. For mathematical simplicity, it is best to consider a design of generator very different from that actually used in engineering practice.

We suppose a small coil be be rotating in a magnetic field. The direction of the magnetic field may be represented by means of an arrow, and the strength of the magnetic field may be represented by the length of this arrow. In Fig. 27, OA is an arrow, representing the magnetic field. This arrow, OA, takes the place of the line OA (joining a fixed point O to a bead A). 'Turning OA' means that we change the direction of the magnetic field. 'Stretching OA' means that we make the field stronger. Both operations can easily be done if the magnetic field is produced by electromagnets, mounted on a bar which can rotate about the fixed point O.

We may take as a standard situation that in which the electromagnets are at E and F and a current of 1 ampere is flowing through them. The operation **a** is to mean that the current in the electro-magnets is increased until the magnetic field at the centre, O, is a times as strong as before.

The operation **ib** would mean that we start from the standard position, make the field b times its standard strength, and then turn through a right angle. The coils would then be in the positions, shown dotted, at G and H, and the magnetic field would be represented by the dotted arrow.

The operation **a+ib** can be interpreted by supposing we combine the two arrangements. We have coils at E and F, through which there flows a current sufficient to produce a magnetic field of a units at O, and *at the same time* we have coils at G and H with a current strong enough to produce a field of b units at the centre. The *combined effect* will be to produce a magnetic field in a direction which lies between the directions OF and OH. The precise position of the arrow representing the combined effect is given by exactly the same rule as before – 'end-to-end addition', more commonly known as the Parallelogram Law, or the Triangle of Forces.

It will be clear to electricians that the arrow representing the magnetic field can be brought to *any* desired position by suitable choice of the size (and direction) of the currents in the circuits $E-F$ and $G-H$. That is, any operation consisting of 'a turn and a stretch' can be put in the form **a+ib**.

To avoid complicating the figure, the small coil rotating about O has not been drawn. The changes in the direction and strength of the magnetic field will, of course, produce corresponding changes in the phase and amplitude of the alternating current generated. It is natural that the operator i should be used in connexion with alternating currents, to show the changes produced by including extra resistance, inductance, etc. What we really do, in using the sumbol i, is to compare the effect of such changes in the circuit with the effect of certain changes (represented by signs such as **a+bi**) made *inside* the generator which produced the current.

The Further Study of i

One question which has been raised in this chapter, but not yet answered, is how to define **e^x**, when **x** is a complex number. **e^x**, in heavy type, is a new label, and we could (if we chose) attach this label to any operation whatever. But this would be very misleading: we should always have to remember that the operation chosen had nothing whatever to do with the ordinary e^x, and we should always be liable to make mistakes through forgetting the difference. On practical grounds it is clearly best not to use the label **e^x** at all, unless we can find some operation which has properties very similar to those of e^x.

We have already found operations whose claim to the labels of **x**, **x^2**, **x^3**, etc., has been admitted, and we know that e^x can be expressed by means of a series containing x, x^2, x^3, etc. It would therefore be natural to form the corresponding series, in heavy type, and to define **e^x** by saying:

$$\mathbf{e^x} = 1 + \mathbf{x} + \tfrac{1}{2}\mathbf{x^2} + \tfrac{1}{6}\mathbf{x^3} + \tfrac{1}{24}\mathbf{x^4} + \tfrac{1}{120}\mathbf{x^5} + \ldots$$

This definition is, in fact, quite satisfactory. It gives us a meaning for **e^x** which can be proved to have all the properties of the ordinary e^x.

It is important to understand what is implied in this definition. If, for example, we wished to find **e^{2+3i}** by means of this series, we should have to replace **x** by **2+3i**, giving:

$$\mathbf{e^{2+3i}} = 1 + (\mathbf{2+3i}) + \tfrac{1}{2}(\mathbf{2+3i})^2 + \tfrac{1}{6}(\mathbf{2+3i})^3 + \text{ etc.}$$

We should then have to work out **(2+3i)^2**, **(2+3i)^3**, etc., and put the answers in. $(2+3i)^2$ turns out to be $-5+12i$, $(2+3i)^3$ is $-46+9i$, and so on. Taking into account the numbers $\frac{1}{2}$, $\frac{1}{6}$, etc. which occur in the series, we thus find:

$$e^{2+3i}=1+(2+3i)+(-2\tfrac{1}{2}+6i)+(-7\tfrac{2}{3}+1\tfrac{1}{2}i)+ \ldots$$

We may now collect together the terms which contain i, and those which are free from i, so that:

$$e^{2+3i}=1+2-2\tfrac{1}{2}-7\tfrac{2}{3} \ldots$$
$$+i(3+6+1\tfrac{1}{2}) \ldots$$

This step is justified, because i can be treated just like x in ordinary algebra, as we noted earlier.

But this result will be no use *unless* we find that the series $1+2-2\tfrac{1}{2}-7\tfrac{2}{3} \ldots$ and the series $3+6+1\tfrac{1}{2} \ldots$ settle down to steady values, when sufficiently many terms are taken. Nor will these series be any use if they turn out to be 'dangerous series' of the type described in Chapter 14.

Actually, these two series are very tame and reliable. The later terms in the series for e^x contain numbers such as $\frac{1}{24}$, $\frac{1}{120}$, $\frac{1}{720}$, which rapidly become very small; and at a certain point, the later terms make hardly any difference to the sum of the series. The rule by which the numbers in e^x are formed is the following: $6=1\times2\times3$, $24=1\times2\times3\times4$, and so on. 120 is 5 times 24. 720 is 6 times 120. The farther we go, the more rapidly do the terms of the series decrease.

It can be proved that the series for e^x is all right (in professional language, is *convergent*) whatever x may be. That is, if $x=a+ib$, it does not matter how large the numbers a and b may be: the series will still converge. If a and b are large numbers, we may have to take a large number of terms before we get a good estimate of e^{a+ib}: nevertheless, as the series *does* define e^x, we have a logical foundation on which to build.

Actually, to find e^{a+ib}, it is better to proceed as follows. $e^{a+ib}=e^a \cdot e^{ib}=e^a$ (*cos* $b+i$ *sin* b). a and b correspond to ordinary numbers a and b. We can look up e^a, *cos* b and *sin* b in tables. But this procedure is possible only *after* we have proved (by means of the series for e^x) that e^x has all the properties of e^x, so that the steps taken above are justified. If our definition of e^x, by means of the series, is not water-tight, then we cannot trust results obtained from this definition.

Mathematicians are therefore forced to study the convergence of series in which complex numbers occur. They have also studied what meaning can be attached to the sign $\frac{dy}{dx}$ when x and y are complex numbers.

We have already seen that the use of i allows us to show a close connexion between e^x, $sin\ x$, and $cos\ x$, a connexion that is surprising, since e^z appears at first sight very different from $sin\ x$ and $cos\ x$. We have also seen that this connexion is of practical use, as it helps us to see the solution of many problems about sines and cosines.

In the same way, the further study of complex numbers throws light on many problems about ordinary numbers. In fact, the subject of complex numbers is one of the most beautiful and instructive departments of mathematics. It gives one the feeling of having been taken behind the scenes: one sees easily and quickly the reasons for results which had previously seemed quite accidental. It is a subject in which calculation plays a small part: its results frequently take forms which one can *see*, and remember, as one remembers a striking poster. By enabling one to see the inner significance of many practical problems, it is therefore of great value for applied mathematicians.

No one could have foreseen that the study of i would lead to such welcome results, any more than the first men who played with magnets and silk could have foreseen the application of electro-magnetic theory to the invention of wireless. In both cases, it just turned out to be so.

When you first learn to use i, you will suffer from a feeling of strangeness. The subject will seem unreal to you. That is inevitable. Any new subject feels strange at first. When wireless first became popular, people felt it to be strange. But the children who are born today take wireless for granted. If, as a war economy, all wireless were to stop, people would say, 'How strange it is, having no wireless!' But no one had wireless in 1914–18, and no one felt it to be strange. Nothing is either strange or familiar in itself. Anything is strange the first time you meet it: anything is familiar when you have known it five years. The more you use i, the more you will come to feel that i is a natural and reasonable thing. But this feeling can come only gradually.

Complex numbers show pure mathematics at its best. Pure mathematics is the study of method. Given any problem, we want to know the best way of attacking it. Many problems, at first sight hard, become simple only if one can look at the problem from the proper angle, if one can see the problem in its proper setting. It is the job of pure mathematicians to classify problems, to suggest that this problem is essentially similar to that, and likely to yield to a certain type of attack. It may not be in the least obvious that the problems are connected: it is not in the least obvious that the equation $x^2 = -1$ is going to throw light on the question of electric lighting. The less obvious the connexion is the more credit must go to the pure mathematician for discovering it; the harder the problem appears, the greater is the credit due for showing that it is connected with some simpler problem.

Engineers do not need to know more than the most elementary results about complex numbers. The more advanced results are chiefly of interest to professional mathematicians, who are inventing and perfecting new methods, which, when complete, can be used by scientists and practical men. Anyone with a taste for mathematics should try, when as young as possible, to gain some knowledge of complex number theory. Books on the subject have such titles as *The Theory of Complex Variables*, *The Theory of Functions*, etc. Too often, boys at school fail to realize how much mathematics there is to know. Talented boys find themselves ahead of their fellows, and begin to think they have a mastery of mathematics. As a result, they waste their last year at school. The first year at college (for those who are able to go there) they meet the best boys from other schools, and experience a tremendous shock.

INDEX